Advanced Sciences and Technologies for Security Applications

The series Advanced Sciences and Technologies for Security Applications comprises interdisciplinary research covering the theory, foundations and domain-specific topics pertaining to security. Publications within the series are peer-reviewed monographs and edited works in the areas of:

- biological and chemical threat recognition and detection (e.g., biosensors, aerosols, forensics)
- crisis and disaster management
- terrorism
- cyber security and secure information systems (e.g., encryption, optical and photonic systems)
- traditional and non-traditional security
- energy, food and resource security
- economic security and securitization (including associated infrastructures)
- transnational crime
- human security and health security
- social, political and psychological aspects of security
- recognition and identification (e.g., optical imaging, biometrics, authentication and verification)
- smart surveillance systems
- applications of theoretical frameworks and methodologies (e.g., grounded theory, complexity, network sciences, modelling and simulation)

Together, the high-quality contributions to this series provide a cross-disciplinary overview of forefront research endeavours aiming to make the world a safer place.

The editors encourage prospective authors to correspond with them in advance of submitting a manuscript. Submission of manuscripts should be made to the Editor-in-Chief or one of the Editors.

More information about this series at http://www.springer.com/series/5540

J. Martín Ramírez · Luis A. García-Segura
Editors

Cyberspace

Risks and Benefits for Society, Security and Development

With a Prologue by Maj. General Carlos G.L. Medina

Springer

Editors
J. Martín Ramírez
Nebrija University
Madrid
Spain

Luis A. García-Segura
Nebrija University
Madrid
Spain

ISSN 1613-5113 ISSN 2363-9466 (electronic)
Advanced Sciences and Technologies for Security Applications
ISBN 978-3-319-54974-3 ISBN 978-3-319-54975-0 (eBook)
DOI 10.1007/978-3-319-54975-0

Library of Congress Control Number: 2017934869

Printed on acid-free paper

This Springer imprint is published by Springer Nature
The registered company is Springer International Publishing AG
The registered company address is: Gewerbestrasse 11, 6330 Cham, Switzerland

To Robert A. Hinde (1923–2017) for his outstanding contributions to CICA and specially to the Seville Statement on Violence.

The original version of the book frontmatter was revised: Dedication text has been added. The erratum to the book frontmatter is available at 10.1007/978-3-319-54975-0_19

Contents

Introduction

This book focuses on one of the most timely topics at the beginning of the twenty-first century: cyberspace, emphasizing not only on its eventual 'negative' challenge as a threat to security, but also on its positive influence as an efficient tool for defence as well as a welcome new factor for economic and industrial production. We will analyze the cyberspace from different and interdisciplinary prospectives: conceptual and legal, military and socio-civil approaches, cyberdelinquency, cyberintelligence applied to public and private institutions, as well as the nuclear governance. It is a scientific product of a selection of papers presented at the XLI CICA (*Conferencias Internacionales sobre Conflicto y Agresión*), on "Mapping the Cyberspace". It was held in Madrid from June 1–3, 2016, co-organized by the Nebrija-Santander Chair for Analysis and Resolution of Conflicts (Center for Conflict Studies) and the Spanish Pugwash National Group.

The CICA conferences started almost 40 years ago in the early eighties of the last century, gathering scholars and researchers interested on the analysis and discussion of the relationship between brain and aggression, and other close topics, such as violence, terrorism, peace, and conflicts at their different levels. From the internal to the international ones, through an integrated, comprehensive, and interdisciplinary approach that considers both biological and psycho-socio-cultural factors. The main characteristic of these CICA meetings throughout the world is precisely this comprehensive approach, crossing disciplinary boundaries. Consequently, it is open to plenty of disciplines: individual and social psychology, psychiatry, physiology, sociology, anthropology, animal behavior, criminology, international law, political science, pharmacology, child development, education, security studies and international relations, law and world affairs, military and peace studies, as well as policymakers.

The academic environment has witnessed a conceptual expansion, broadening out from issues of traditional security and military strategy to include conflict transformation, human security, peace building and governance. There is now much greater awareness that peace and conflict depend on a vast range of factors, such as inter alia inequality, human rights, arms control, international norms, and psychological and mobilization processes. An expression of this scientific concern was our specific approach to security and defense, which led to the organization of a previous XL CICA that was mainly focused on the "Protection of Critical Infrastructures", one of the twelve main threats mentioned by the Spanish government within the frame of its current strategy of security and defence (*Estrategia de Seguridad Nacional* and the *Directiva de Defensa Nacional*, 2013): armed conflicts, terrorism, cyberthreats, organized crime, economic instability, energy vulnerability, flow of irregular migrants; weapons of mass destruction (WMD), espionage, natural emergencies and disasters, vulnerability of the maritime space, and threats to the critical infrastructures and essential services. Their collapse would generate an alert situation all over the country. This explains the importance of minimizing to the *maximum* their risks: the prevention of these kinds of threats has become one of the priorities in the framework of the security of any country. The main results were published in two academic books: *Security in Infrastructures* (Cambridge Scholars Publishing) and *Retos de la Seguridad* (Thompson Reuters).

The scientific fruit of the previous forty CICAs held to date at 16 countries in five continents (Spain, Chile, Colombia, South Africa, Sri Lanka, USA [California, New England, and Georgia], Greece, Zambia, Italy, England, Northern Ireland, Mexico, Poland, Turkey, Hungary, and Bulgaria), is reflected in the elaboration of 30 odd publications, most of them in English language (see: http://www.cicainternational.org).

Following the same perspective, the 41st CICA chose as its main goal another of the 12 threats mentioned by the 2010 Strategic Concept for the Defense and Security of the Members of NATO, as well as by the Spanish government: *cyberthreats*. More specifically, it focused on the current knowledge and research on the cyberspace, a topic quite close to the Pugwash Movement, Nobel Peace Prize in 1996, and co-organizer of the conference. Participants were from 22 countries from five different continents: China, Iran, Nepal, Qatar, Russia, Palestine, Poland, Germany, Croatia, England, Wales, Italy, Canada, USA, Nigeria, South Africa, Colombia, Cuba, Dominican Republic, Venezuela, and Spain, among others.

More specifically, we had the active participation of reputed scholars from different continents and from different fields of expertise as distinguished guest speakers: the Chairmen of the Pugwash Movement in Germany, Canada, Croacia, and Spain, the Chairman of the Missile Technology Control Regime (MTCR), the Director of the Russian Center for Euro-Atlantic Security, former Directors of the EU Satellite Center, of NATO Staff for Cooperation and Regional Security, and the CIS Division Spanish Navy HQ, the President of the Colombian Institute of Studies on Psychosocialbiology and Intervention in Violence, as well as the Attorney Coordinator on Cybercriminality in Spain, representatives of the Spanish National Police and Guardia Civil and of some private institutions closely related to cybersecurity. The Spanish Joint Ciberdefence Command (MCCD) covered a symposium on the military approach to the topic and, last but not least, the keynote speech was given by the Advisor to Vice President of Iran and Head of Iranian Atomic Energy Organization.

This present volume on impact of cyberspace brings together a collection of scholarly works, authored by 24 international researchers and leading thinkers representing nine countries from Europe, Africa, America and the Middle East, addressing quite relevant issues on cyberspace in general and, more specifically, on cyberdelinquency, and on cyberwarfare, from an interdisciplinary perspective. The academic background of its contributors is quite diverse: militia and police, law, diplomacy, aggression and conflict studies, and psychology. This approach encourages a broader perspective and thought process, trans-discipline and global collaboration and cooperation, and an integrated synthesis of knowledge. This eclectic cast of authors approaches the main topic from three different points of view:

The first part opens with six chapters focused on some interesting narrative and conceptual considerations on cyberspace. The first by Elizalde, a Spanish theoretical physicist, offers a personal accounting on how the cyberspace arose; Harries, a former FC Canadian officer dedicates the second one to narrate the context and consequences of its mapping; the next two chapters are dedicated to some conceptual and legal considerations in general, by Cortijo, a former Spanish Police Commissioner, and applied to some Latin American developing countries, by two law scholars, the Dominican García-Segura and Cayón, from Spain; Gago, a Spanish economist and experienced businessman, assesses its effect on trade and business in the fifth; and the final chapter, by Farnicka, a Polish sociopsychologist, analyzes its psychological impact.

Another six chapters written by Spanish specialists, plus a final one by a British scholar, compose the second part, on cybersecurity. Two scholars, Carpintero and Otero, present cyberspace as a platform for organized crime, in the first chapter; Ramirez, expert on conflict and aggression, adds some criminal aspects of cybersecurity in the second; another two authors working in private companies on cybersecurity, explain their experiences on managed services, Velasco, and private investigation, Cesteros; the fifth and the sixth chapters are focused on some psychosocial considerations, by Fernández-Rodríguez and Miralles, and the use of cyberspace for terrorist purposes, by Payá and Delgado. The section is closed by a quite provocative perspective on the insecurity and governance in cyberspace, by Wright.

The third part discusses directly the cyberwarfare, in five chapters: the first two, written by a Palestinian jurist, Boulos, and by two members of the Spanish Defence Forces, Col. Domínguez and Admiral Bauzá, look specifically at the Tallinn Manual; in the third, the Italian ambassador Trezza analyzes the options of diplomatic negotiations; the Iranian ambassador Soltanieh and his co-worker Esmailbagi focus the fourth chapter on the security of nuclear facilities; and finally, the Swedish scholar Westberg suggests a possible positive value of the existence of cyberwarfare, as an efficient tool for preventing wars.

Cyberspace is completed by a Prologue by General Medina, Chief of the Spanish Joint Cyber Command, as well as by an Introduction and an Epilogue by both co-editors.

In sum, this book intends to transmit an integrating and synthesizing approach to cyberspace to scholars of security and international relations, law and world affairs, military and peace studies, as well as to policymakers, and to the general public who are interested in keeping up with this new global area of concern. We hope its reading may help to achieve our purpose of leading to better knowledge and broader understanding of this "fifth domain".

December 2016

J. Martín Ramírez
Universidad Antonio de Nebrija
Madrid, Spain

Prologue

Every day, we spend more and more time in cyberspace and, in many cases, solving problems that have originated in the real world. For example, this morning we may have been woken up by our cell phone, which takes the real time from the Internet, written and answered many emails, paid attention to social networks, groups of "WhatsApp", and so forth.

How did we live a few years ago, when we did not have email, cell phone, or the World Wide Web? It's hard to remember. Cyberspace has changed the way we connect with others, with individuals, with companies, or with the Government. It has brought globalization to a truly global dimension and it has practically erased the relevance of having different schedules or the time zone differences in the world. From this point of view, probably the impact of cyberspace on mankind today is greater than the invention of printing in the fifteenth century.

But what do we know about cyberspace? What do we know about that place where we are spending more and more time and is affecting our lives more and more? Not very much. Our perception is that it is a kind of mystery that is, somehow, behind our personal computer or our mobile devices and only few technically qualified people know it in depth. It is thought that cyberspace has a virtual nature; however, this is not exactly true. Cyberspace consists of absolutely tangible elements such as personal computers, mobile devices, servers, routers, fiber optic cables, satellites, operating systems, applications, etc.

As so many other things, cyberspace is neither good nor bad by itself; everything depends on how you use it. We must recognize that it has been created and developed only with its functionality and with the advantages it brings to its users in mind. However, until recently, its security has not been thought about. Its birth and vertiginous evolution have been guided by the idea of freedom and the absence of borders, limitations or any regulation.

Given these circumstances, it was only a matter of time for the cybercrime to appear. This has involved a progressively changing format: from the isolated delinquent to the organized criminal group, increasing the transcendence of cyberattack at breakneck speed. We know that today's society as a whole is threatened when the objective of cyberattack is critical infrastructure or vital services, essential for the undertaking of our daily activities. If these are not protected enough, they become fragile and vulnerable; and these cyberspace vulnerabilities can be exploited by terrorist groups, and even by states.

In order to counter these threats, the concept of cybersecurity was born. It is about including means (hardware and software) and procedures to reduce the risk of successful cyberattacks. However, cybersecurity begins with the awareness of users. If we do not feel really threatened, we will hardly foster cybersecurity.

And on the military realm? Can we talk about a real cyberwar? Military operations also rely heavily on cyberspace. Modern weapon systems provide their authentic capability when interconnected. Without access to cyberspace, the Western Armed Forces would revert back to that of the past. As a matter of fact, cyberspace has been added to the so-called "conventional" domains (land, sea, air and space) to become the fifth domain of operations. Military operations can be planned, conducted and executed in cyberspace, as NATO has recognized at the last summit of Heads of State and Government held in Warsaw in July 2016.

It is hard for an armed conflict to be solved exclusively with cybernetic actions, but it is beyond doubt the main role that these actions currently have in any military conflict.

"Cyberspace" contributes to improve the knowledge of cyberspace and provides different points of view on this issue.

The first part focuses on cyberspace, analyzing its essence, the consequences of its existence, its effect on trade, business and underdeveloped countries, its legal aspects (security vs. freedom quandary) and the impact on the safety of individuals and groups.

The second part deals with cybersecurity, identifying the link between crime and cyberspace, analyzing its evolution, the possible collaboration of private investigation with institutions, the psychological characteristics of cybercrime and the use of cyberspace for terrorist purposes.

The third part is devoted to the cyberwar and its legal facet: the Tallinn Manual, the possibility of the application of the law of armed conflicts to cyberwarfare, the negotiation options, the risks that cyberspace can pose to nuclear facilities and the feasibility that engagements in the fifth domain can avoid conventional wars.

Because of the breadth and variety of the aspects discussed, "Cyberspace" is interesting for all kind of readers, both for experts in this field and for those who are just curious to know more about this new environment in which, every day, more and more often, we place our interests.

Maj. General Carlos G.L. Medina
Chief of the Spanish
Joint Cyber Command

Part I
Cyberspace

On How the Cyberspace Arose to Fulfill Theoretical Physicists' Needs and Eventually Changed the World: Personal Recallings and a Practitioner's Perspective

Emilio Elizalde

Abstract The very humble origins of the Cyberspace, and all the related developments that smoothly conspired and converged towards this concept, making its emergence possible, as the personal computer, TEX and LATEX, the Fax, the internet, the cellphone, and the World Wide Web, are discussed, always from a personal perspective. A separate, comprehensive explanation of the reasons for the appearance and sub-sequent evolution of each of these different phenomena, with explicit assessments and a future prospective of the main subject, is presented.

Keywords Cyberspace · Theoretical physics

1 Introduction

Let me start from the beginning. I am a theoretical physicist and a cosmologist, as well as a mathematician (Elizalde 2016). My main research fields are zeta function regularization (Elizalde 2012; Elizalde et al. 1994), dark energy issues (Elizalde et al. 2004; Cognola et al. 2008), mainly in relation with quantum vacuum fluctuations (Bordag et al. 1996), and modified gravity (Elizalde et al. 2003). In 1982, during my second visit to the II. Institut für Theoretische Physik and the Deutsches Elektronen-Synchrotron (DESY), in Hamburg, as a Humboldt Fellow (AvHumboldtF 2016; AvHumboldtE 2016), I came to know about the TEX project, a typesetting system developed by Knuth (1969), and existing since 1978. It was a wonderful, very ambitious project. A couple of years later, back at the Department

Submitted: 22.10.2016; Accepted: 2.11.2016.

E. Elizalde (✉)
National Higher Research Council of Spain, Instituto de Ciencias del Espacio (ICE/CSIC and IEEC), Campus Universitat Autònoma de Barcelona, Barcelona, Spain
e-mail: elizalde@ieec.uab.es

J.M. Ramírez and L.A. García-Segura (eds.), *Cyberspace*, Advanced Sciences and Technologies for Security Applications, DOI 10.1007/978-3-319-54975-0_1

of Theoretical Physics of Barcelona University, as an Assistant Professor, I was chosen to be the responsible for the preprint interchange with other universities worldwide. Sure, this had always been a very important issue for theoretical physicists anywhere, right then and also in the past. The reason was the following. When one would send a paper for publication to a specialized journal, one would always have to wait for several months (sometimes one or two years) before the paper managed to make its way to a publication [even today this is a serious issue (Powell 2016)]. In a very quickly developing field, as the one in question, this was at the time a really dramatic drawback. In other words, the importance of a rapid dissemination of our work, under the form of the so-called preprint (Preprint 2016), was paramount. But the whole procedure was extremely cumbersome, as I could appreciate (suffer is a better word), first hand, during the period that I was in charge of this business for my Department. I will now try to describe what was going on.

To start with, we had to do a careful selection of a certain number of the best Departments of Physics in the World, being the very first step to decide on this precise number; what depended on the money available to carry out the full task. At the epoch I was in charge, we selected in Barcelona over 200 universities (250 was soon to become our all-time record), with annual revision of the list. We were bound to choose the most interesting places where to send the preprints, both because we assured in this way that our work would get to be known by the top specialists in our research fields and-equally or even more important, because there was an (unwritten) law of reciprocity, by which, on doing so, we would also get the preprints issued by those departments where the best of our colleagues were working at that time. I can assure you, choosing the list was not always that simple (Inspire HEP 2014).

The second step, after completing some valuable, original research project (this was actually our real duty, of course, aside from giving lectures), was to type our preprints using our precious typewriter (sometimes there was a line to be kept for that), taking care to allow enough space for the special characters, formulas and equations, which we had to fill in later, always by hand. If something got wrong during the typing, or if we discovered some mistake, we had to start over, sometimes right from the beginning! Each month or two I would collect all preprints written in our group and, with the help of a couple of post-docs we would make the, say 220 copies of each of the say 8 papers; what then meant classifying the pages, stapling each copy together, writing the envelopes with the 220 addresses, filling each one of the envelopes with the 8 copies, closing and sealing the 220 envelopes, bringing the whole package to the post and, finally, paying good money for the expedition of the package.

Even now, by just recalling and describing to you in detail the whole process we had to go through I get again pretty tired. The whole thing had a considerable cost, both in terms of personal effort and money (and also ecological! although this concept just did not exist right then). You should keep in mind, and perhaps read once more, the several paragraphs above in order to better grasp the enormous importance it had for us the developments that were about going to take place, almost simultaneously and on several parallel, complementary levels, and which eventually conspired together to give birth to the most important social revolution, at planetary scale, in human history.

2 A Most Humble Aim: Producing Professionally-Looking Preprints and Sending Their Encoded Files by Using Computers

The first remarkable development, as already advanced in the Introduction, was the appearance of TEX (Knuth 1969), and a few years later, in the middle 80s, of LATEX (Leslie Lamport's TEX) (Lamport 1986), the poor brother of TEX but much more simple to use and thus more appropriate for the average theoretical physicist (or mathematician). LATEX (even more, TEX) was (and still is) an incredibly powerful tool. It permitted you to write absolutely perfect, professionally looking papers, allowing anybody to produce high-quality articles and books using fairly minimal effort. And that under the form of a binary file, which could be sent away without the least distortion, not even of a single bit (lossless, as we would say today: it would yield exactly the same final result on all kind of computers, at any place in the world and point in time.

Needless to say, typewriters had in the meanwhile (in just very few years) become fully obsolete with the advent of personal computers: Atari, Altair, Spectrum, Amstrad, IBM, to name only a few of the most significant ones, in terms of popularity (Knight 2014). The marvelous and very modern IBM ball typewriters I had used to type my papers when I had been at DESY for the first time (1979–80) were suddenly of no use any more, at least for us, theoretical physicists (although they were still employed, for a while, for writing down ordinary forms or filling up official documents).

But this immediately leads us to the second extraordinary development (in fact, the first one in order of importance, even more with regard to the title of the present book). To wit, the LATEX file encoding all information of the paper (or book, given the case) had to be transmitted, e.g. distributed, as with previous, paper preprints was done in the way I have described before. However, this was not going to take place, now, by using the pony express or any other postal services anymore. In no way by sending again, inside of an envelope, the magnetic tapes or discs or any kind of material support containing the files. The groundbreaking idea was now to connect computers to computers and transmit the files immaterially, using this connection: the internet. According to the Wikipedia (2016) presently "the Internet is the global system of interconnected computer networks that use the Internet protocol suite (TCP/IP [17]) to link billions of devices worldwide. It is a network of networks that consists of millions of private, public, academic, business, and government networks of local to global scope, linked by a broad array of electronic, wireless, and optical networking technologies," which is a very nice definition, indeed.

In the middle eighties of the past century, aside from an extensive military network already in place in the USA, the Arpanet (existing since the 1960s, and which was in fact the origin of the concept of the present internet itself), 'the international internet' consisted in a connection of just *three* computers: one in Los Alamos National Laboratory (LANL), USA, one in the European Organization for Nuclear Research (CERN), in Geneva, Switzerland, and one in DESY, in Hamburg, Germany. Actually, these three institutions had already been playing, for many years, a very important role for us all, theoretical physicists: they maintained archives where all preprints they received where classified by subject, given a number, and properly marked with a reception date. Those were considered by everybody to be very reliable 'oficial' archives; for instance, in order to establish priority of a certain result or discovery. Of course this had been always established by the date of the corresponding publication, finally, in an international journal. But in such a fast developing area no one would wait until publication, to establish such priority: a clever idea or result could be stolen [and some certainly are (Southwick 2012)], or at the very least much delayed, during the long revision process that led to publication.

With the use of these archives, priority was established, in fact, since the very moment when your work got registered in any of those repositories (better in more than one!) under the form of an archived preprint with a reception date. Just because these archives were reliable, public, independent and fully transparent to the whole international community. This issue was actually of paramount important, in particular, for a young scientist as I myself was at that time. One always had the highest expectations and believed that the work just finished was to become a masterpiece, or, at the very least, a very much influential paper.

With the new advances, LANL, the *Conseil Européen pour la Recherche Nucléaire* (CERN), and DESY could actively share their collections of preprints and start to work together. The first step was taken to provide open access, from the distance, to the electronic preprint collections by means of the internet computer connections already mentioned. Thus, in DESY there was a list where you would enter your name, and a date and time slot to be filled, which procured you half an hour of use of the computer, which you then would spend by looking from Hamburg's computer at the preprint lists at CERN and Los Alamos. You then could write down the references you were interested in and ask later the authors of the papers to send a copy to your address. It took still some time before we were able to retrieve the corresponding file through the computer connection (for it was quite slow and of poor bandwidth). Needless to say, all enquires where then done by entering your computer orders line by line, being UNIX the computer language, and using the whole screen at a time (the Windows concept had still to appear).

Now a revealing anecdote. That same summer my two sons, both music students, who were staying with me and my wife at the DESY Hostel for the period, needed to practice on a piano during their long vacations. At DESY the only piano available was the one in the main Auditorium. It is still there, at the same right corner of the stage, and that after 30 years, as I could check last month (Desy 2016) when I visited DESY again, to take part in the Annual Conference and in the

Colloquium *Local Quantum Physics and beyond*, in memoriam Rudolf Haag (by the way, a renowned professor, who had been my host during my stay in Hamburg as a Humboldt Fellow, in the 1980s). Curiously enough, the procedure I had to follow in order to reserve the piano for my sons to play, and the one for reserving the computer for me to use, as explained, were exactly the same! I would bet that, with the piano, you still have to go right now through the same procedure (the only possible improvement being that you will now enter your appointment at the piano list conveniently using your cellular). This makes me ponder how enormously things have changed, during these same 30 years, concerning the uses and possibilities of computers. In particular, in relation with the issue discussed in this paragraph: entering a list to reserve computer time to connect somewhere (…?) seems a ridiculous concept, of the Stone Age! It is amazing!

3 Paul Ginsparg and the ArXiv Concept

Another important step was going to be taken soon. The ArXiv, a repository of electronic preprints or *e-prints*-which was the name soon given to the LATEX files I was talking about-was put in place on August 14, 1991. As clearly explained in the Wikipedia (WikiArxiv 2016), it was made possible by the low bandwidth TEX file format, which allowed scientific papers to be easily transmitted over the incipient internet. In fact, according to the Wikipedia, this had been started a bit before, around 1990, by Joanne Cohn, who had begun e-mailing physics preprints to colleagues as TEX files, but the number of papers being sent soon surpassed the very limited capacity of mailboxes. Ginsparg (2016), a theoretical physicist who had begun as a junior fellow at Harvard and had moved as a member of staff to LANL, quickly understood the need for a capable central storage device, and in August 1991 he finally materialized the idea to install a central repository mailbox to be accessed from all computers with an internet connection. He was very proud that he could store so many preprints in his own, not so big computer, and I can certify that the service he provided to the international community of theoretical physicists, for the first few years at least, with such limited resources, was incredible.

Important additional access modes were added soon, namely the File Transfer Protocol (FTP) in 1991, Gopher in 1992, and in 1993 the World Wide Web (WWW). I distinctly remember all these steps, as very welcome additions to the process, constantly improving preprint production, file transmission, and access to the repository. The ArXiv began as a theoretical physics archive only: the LANL preprint archive, its domain name being xxx.lanl.gov. Soon other disciplines of physics were added, and then astronomy, mathematics, computer science, nonlinear science, quantitative biology, statistics and, lastly, quantitative finance. Ginsparg moved to Cornell University in 1999 and he changed the name of the repository to arXiv.org. Presently, it is hosted by Cornell, but it has 8 mirrors scattered around the world. In 2008 the ArXiv reached the half-million article mark, and the one million milestone in 2014. The submission rate approaches the 10,000 papers per month, right now. Not just the volume, but the quality and importance of the ArXiv have been increasing constantly and, in a way, the whole process has culminated in the most relevant publishing movement of our days, known as Open Access.

At a point, the registration of a preprint in the ArXiv has been given equal status to that of a paper regularly published in a most prestigious specialized journal. For one, let me just recall the famous case of the Russian mathematician Perelman (2014), who proved the Poincaré conjecture (ClayPoinc 2016a), the first (and the only one till now) of the Seven Millennium Problems (Milleniump 2016) to have been solved. Perelman issued his solution in a series of three e-prints that appeared in the ArXiv between Nov. 2002 and Jul. 2003 (Perelman 2002, 2003a, b), but which he refused to send to a journal for publication, with the argument that this was fully unnecessary, given the universality of the ArXiv. He went on saying that everybody in the World could check if his results were right-of what he was completely sure and was, in fact, definitely proven to be the case. Perelman's proof of the Poincaré conjecture is considered to be one of the most important achievements of the past Century.

Although, in order to meet the criteria to receive the one-million-dollar Clay Millennium Prize (ClayPoinc 2016b) for this discovery, it was compulsory that the proof would be published in a regular journal, the Clay Foundation gave finally up, and offcially acknowledged (for the first time ever) that Perelman's e-prints registered in the ArXiv were fully equivalent to regular journal publications (for the reasons already given). This was the culmination of the idea, I had advanced in a previous section, that priority of a scientific result is now fully, legally established in terms of its acceptance and registration as an e-print in the ArXiv.

The above case has also prompted many other colleagues, some of them quite distinguished, to do the same as Perelman, at least from time to time: the number of important results that have only been issued as e-prints, never having been published in a regular journal, is constantly increasing-and some of them have accumulated record numbers of citations (Lariviere et al. 2013).

4 The World Wide Web

Scholars generally agree, as well described in the Wikipedia, that a turning point for the WWW began with the introduction of the Mosaic web browser in 1993, a graphical browser that had been insistently demanded by our community of theoreticians in order to deal with figures and graphs (I will discuss this issue in much more detail later). Before the appearance of Mosaic, graphics could not be easily incorporated in web pages, together with text. As a consequence, the Web was not so popular as other protocols, as e.g., Gopher. The graphical user interface of Mosaic was decisive in making of the Web, by far, the most popular internet protocol. In October 1994, the physicist Tim Berners-Lee founded the World Wide Web Consortium (W3C) at the Massachusetts Institute of Technology (MIT), what was the culmination of this process. But let us proceed slowly and recall in detail the whole story.

Soon after graduation, Berners-Lee had become a telecommunications engineer and after working for the industry he had been a fellow at CERN, in Geneva (Switzerland), for ten years, where he first became acquainted with the needs of the huge community of physicists working there and then devoted his best efforts to satisfy them. We should not forget that the CERN, already mentioned in previous sections, is now the greatest temple of Theoretical and High Energy Physics in the World. It has the most powerful particle accelerator, by far, namely the Large Hadron Collider (LHC). Experiments in the LHC have allowed to reproduce and study in all detail the same physical conditions of our Universe when its age was just of some three femtoseconds, that is, when just 3×10^{15} s had elapsed since the Big Bang singularity.

CERN was already at the time the largest Internet node in Europe, and Berners-Lee clearly saw an opportunity there, as he personally explains:

> I just had to take the hypertext idea and connect it to the transmission control protocol (TCP) and domain name system (DNS) ideas and ta-da! the World Wide Web…Creating the web was really an act of desperation, because the situation without it was very difficult when I was working at CERN later. Most of the technology involved in the web, like the hypertext, like the Internet, multifont text objects, had all been designed already. I just had to put them together. It was a step of generalizing, going to a higher level of abstraction, thinking about all the documentation systems out there as being possibly part of a larger imaginary documentation system (Berners-Lee 2016).

However, as advanced, the Web was actually founded when Berners-Lee had already left CERN, at another World leading institution: The Massachusetts Institute of Technology Laboratory for Computer Science (MIT/LCS) with support from the Defense Advanced Research Projects Agency (DARPA), which had pioneered the Internet, as already explained above. A second site was founded just one year later, at the French laboratory *Institut National de Recherche en Informatique et en Automatique* (INRIA), with support from the European Commission Digital Information Society (DG InfSo, recently renamed DG Connect). A third continental site was then created at Keio University, in Japan, in 1996. Although in the middle 1990s the number of websites remained still relatively small, many of the best known sites were already functioning and providing popular services.

Probably the most important idea of Berners-Lee, while at CERN, was to join together the hypertext and the internet. In doing so, he developed three essential technologies, which sound very familiar now: (i) a system of global, unique identifiers for all resources on the Web, first known as the Universal Document Identifier (UDI), and later specified as the Uniform Resource Locator (URL), on one hand, and the Uniform Resource Identifier (URI), on the other; (ii) the publishing language HyperText Markup Language (HTML); and (iii) the HyperText Transfer Protocol (HTTP).

The Web had a number of differences with respect to the other available hypertext systems, in that it required only unidirectional links, and not bidirectional ones, as the rest. This fact rendered it possible, for anybody, to establish a link to another computer without the need to have to wait for any action by the responsible of the second resource. It simplified very much the process of implementation of web servers and browsers. As soon as 30 April 1993, CERN officially announced, that the WWW would be free to anyone (Smith and Flückiger 2016). This fact that the Web was non-proprietary was decisive, since it made possible to develop new servers and clients and to add extensions without having to ask for licenses all the time. As the use of the Gopher competing protocol was not free, this announcement immediately produced a decisive movement of clients towards the Web. Berners-Lee had made his brilliant idea freely available, with no patent to be filled and no royalties to be paid. The WWW Consortium also decided, on its turn, that Web standards should be based on royalty-free technology, so that they could be easily adopted by everybody in the World. This fact contributed to the extremely quick and successful expansion of the Web in the whole World since the middle 1990s.

The impact of the Web on the world society has been enormous since then, but not so specially on theoretical physicists, mathematicians or theoreticians of any other kind, as I will now explain below. The first web site was opened at CERN, and had the address: info.cern.ch. The first web page http://info.cern.ch/hypertext/WWW/TheProject.html contained information on the WWW project. On Aug. 6, 1991 it was set online and, as the project was quickly improving, changes to the page were made each day. In a list of "80 cultural moments that shaped the world," chosen by a panel of 25 eminent scientists, academics, writers and world leaders, the WWW has been ranked as number one of all them, for being "The fastest growing communications medium of all time, the Internet has changed the shape of modern life forever. We can connect with each other instantly, all over the world (British Council 2014)".

This development was absolutely unforeseen by the creators of the Web and by the responsible at CERN and MIT. No one could guess that by just trying to implement, to bring to its ultimate perfection, the idea of dealing with all aspects of scientific work production, sharing, cooperation and distribution, by then going further to include figures, images and even little movies, necessary to better describe some important scientific results, would be later used to satisfy so many different needs of all areas of knowledge and, far outside from that, of the common human being and of the universal society. And that it was going to fulfill dreams no one had even imagined to be able to dream in the middle 1990s. In the following sections I will further elaborate around this issue.

5 Fulfilling the Needs of Theoretical Physicists

When I now compare what I, as a theoretical cosmologist or mathematician, have gained from the recent web developments, say since day one of the present Millennium, my answer must be: fairly little. The needs of an average theoretician or high-energy physicist, which were at the very origin and conception of the humble initial proposal, which culminated in the Cyberspace, had been totally fulfilled by the end of the last Millennium, even before, probably. I will put a specific example in order to justify and properly explain this claim. During my career as a physicist, I have had a number of Ph.D. students and I will now just compare their daily work and behavior, during the preparation and finalization of their respective thesis works and until they got a post-doc.

My firsts students, in the middle to late 1980s used typewriters for writing down the text part of their manuscripts, filling later formulas, equations and special characters by hand. They customarily attended international schools and conferences and were much aware of the latest results in their research subjects through preprints that they solicited by postcard, normally to the authors themselves or, alternatively, to the already mentioned repositories. At that time, if they were lucky, they could manage that some preprint was sent to them by Fax (what had some associated cost that I would cover through my research funds); this was indeed another precious possibility at the time, not mentioned before, which I will consider later in further detail, and in a more general context. Although they also consulted the preprints displayed on an expository shelf at our Department of Theoretical Physics (University of Barcelona, UB), arriving from LANL, CERN, DESY and several dozen other universities (see Sect. 1).

The appearance of LATEX, which we got in 1987, was enthusiastically welcomed in our Department, but not so by other groups of the Physics Faculty, who remained quite indifferent or were unwilling to lose even the least amount of time in that matter. Actually my then student Enrique Gaztañaga and myself were the very first in the University of Barcelona (UB) to get the LATEX program; I clearly remember the day when we went to visit the group of Theoretical Physics of the Autonomous University of Barcelona (UAB)—which was the first group to buy the program in our local community-to get hold of a copy. Soon we started, with Enrique, to write preprints in LATEX; this was certainly a revolution and, needless to say, he wrote his Ph.D. thesis in this code. Compiling each page took a couple of minutes at the time, but the final result was astonishingly professional. And you could improve the manuscript so easily, spotting the mistake and changing only this little part. Sure, you cannot experience such a wonderful feeling in all its depth if you did not live through this period, if you were not actually there. My student August Romeo was also born to theoretical physics at that time. Both August and Enrique soon mastered in LATEX and I learned a lot from them.

I have already described the fast evolution and improvements that took place immediately afterwards. In 1991 the ArXiv was born and there was no more need to send postcards in order to get interesting preprints, nor had we to wait for them to be displayed on our preprint shelf on Friday morning every week. Of course, some improvements continued to happen in the 1990s, as in the searching, classification, speed in sending and getting e-prints, compilation got much faster, etc. But from the end of the 1990s onward the advances, as what concerns our field of research, have been rather cosmetic or superficial.

When I now do compare how my students from the late 1990s and my present students work, the differences are no more that remarkable. They regularly access the ArXiv (now from any place, that is true, including their mobile phones) to easily search for information about the subject of their work. They write papers in LATEX and send them also to the ArXiv, on their turn. However, I would bet that (at least on the average) present day theoretical physics students do not master LATEX to the same level as we, little experts, used to do in the old times. For us it has always remained something very, very special.

Today they have several other alternative possibilities. For one, the text processor Word, which was actually always there (in parallel to LATEX), to write ordinary documents, has become very powerful and allows now to insert beautiful equations, too. And, on the other hand, Word is also very useful in connection with clean and very professional presentations by means of Power Point. Let me say, on passing, that Word and the rest of the Office programs represented for the administrators and the society in general, a parallel revolution as powerful and important (or even much more, according to the number of people influenced by it) as that of TEX for the scientists. My whole point is just that we, theoretical physicists, were the spark, the ones who ignited the fire, and, much more than that, the ones who created the whole firework castle, e.g., the language, the connections, the whole infrastructure that made it possible for the whole thing to explode and expand.

Actually, talking about the Office programs, this is another important issue I have not been dealing with before, namely that of the evolution of the way physics results are presented in seminars, lectures, and conferences of all kind. That is, the revolution that brought us from 'transparencies', always written by hand, and 'retro-projectors'—with the related thousand and-one problems of bulbs that kept fusing all the time-to the ubiquitous use of computers and Power Point (or Acrobat pdf) presentations with animations of all sort trying to captivate the audience at any price. There has been an enormous change in that direction but I would say this one is not of an essential nature, in the sense that the other changes here described have been.

To establish this point on firm grounds and with very few words, let me just mention a couple of examples. Recently, in the spectacular Starmus meeting in Tenerife (Starmus 2016) I could witness (and a dozen Nobel Prize Laureates participating there also) how Sir Roger Penrose gave his awaited for presentation (Penrose 2016). He did it with the help of a computer and other modern means, of course, but it simply consisted of a bunch of projected pdf scans of ordinary handwritten pages. They looked exactly as in the old times! It was a big surprise for the whole audience. But this one is in no way the only such case I have seen in the last half a year (another colleague did the same in a scientific presentation at the Benasque Center of Theoretical Physics in September this year). As another, different example I may add that I still attend, from time to time, to interesting seminars given on the blackboard. And, moreover, it is remarkable that many colleagues often tend to think that these presentations are in fact deeper, more interesting, and that they even captivate the attention from the audience better than many up to date Power Point fashionable flashes.[1]

[1]"Power tends to corrupt, and absolute power corrupts absolutely" as Lord Acton liked to observe. In the present context, we may also say that PPT corrupts absolutely, in the sense that it doesn't have to be used always, but only when it is really needed (Martín Ramírez, personal communication).

Back to the issue under discussion above, my present students still study with the same reference e-books, and also with ordinary books, of course (sometimes coming from pirate sources), exactly as in the last decade of the past Millennium. Maybe the only significant difference is the one just mentioned by passing, that access to information about their working interests has now become much faster, space and timely ubiquitous. And also the existence now of the Wikipedia (not of much use for doing hard and original research work, actually), and of other internet resources, as the extraordinary MIT on-line lectures (Lewin 2016), and those from many other top universities and research labs. But neither of those are really fundamental improvements for the work of a theoretical physicist or mathematician. Our professional needs were reasonably fulfilled, I repeat, before the end of the previous Millennium. No one suspected however (but maybe for a couple of real visionaries) that the revolution put forward by a little bunch of theoretical physicists would be growing and growing without an end (it is still on its way today, faster than ever) to first equal, and subsequently surpass, all previous revolutions in the history of Humanity as the steam machine, the railroad, the telegraph/telephone, or the cinema, or whichever you may right now think of, for its enormous and almost simultaneous impact at world scale that has made, in many respects, of our planet Earth a global village. Its projection to the future, still to come, will be for sure not less impacting. And it even seems now to be not so difficult to predict, as I will try to show in a forthcoming section.

6 Once More the Same Discourse, but from a Different Perspective

I will here start again, from the very beginning, the same story, but now from a completely different perspective. Let me now look at the whole development not from the point of view of a theoretical scientist, but from that of a person with a basic cultural background of any kind. The starting emphasis will be now on the Telefax, usually known as Fax (Fax authority 2016): a system for sending and receiving printed materials, transmitted on the telephone line, normally to a telephone number connected to a printer. The first step is to scan the original page of paper and, simplifying for the sake of brevity, the ordinary perception of the whole process is that the actual sheet of paper is traveling through the telephone line, until it finally appears at the end of the same. For sure, it is not the paper that travels, but the pixelized information of its content, which can be recovered and printed on a blank page of paper at its end.

The FAX was, of course, already in the 1980s, an available possibility to send preprints abroad, and it was in fact being used in some specific cases; but it could not be adopted as a real solution to the problem we had (see the beginning of this Chapter). For many obvious reasons, starting with small bandwidth, long transmission times, and the very high associated costs. However, it was very much employed by the administration, big companies, etc. During a visit that I paid to the Physics Department of Leningrad University in 1989 (Gorbachev's *perestroika* time), in order to initiate scientific relations between our Physics Departments, this was my only reliable, fast, and daily connection to the outside world.

Actually, a predecessor of the Fax concept had been already invented in the 19th Century, although the first commercial version of the modern fax machine was patented in 1964 by Xerox Corporation and got the name of Long Distance Xerography (LDX). No doubt, xerox-copying and printing is a main ingredient in the process. By the late 1970s many companies (mainly Japanese ones) were in the fax market. The way the fax works, the fax technology, has improved considerably with the years. In particular, in the early 1980s Ethernet enabled fax services were already in operation. But, as for now, I just want to concentrate on the idea of the fax machine as a very clever way, more or less sophisticated, to transmit all kinds of printed material, that is both text and images.

Seen from this point of view, the very first steps of the internet revolution aimed at a very similar goal: to transmit on the internet line (in a way quite similar to the telephone line), the LATEX files corresponding to the scientific preprints we were interested in. In a way, there was not much difference of principle between these two conceptions, only technical issues and, of course, associated costs, as explained. The Fax concept has not substantially evolved since that time, although, to be fair, one must consider, separately on its own right, optical scanning, one of the main ingredients of the fax machine, namely the encoding in a file of the information contained in the scanned picture in a compressed way, always with some loss with respect to the original, depending on the nominal resolution of the device we are using. In fact, scanning has become a habitual task in our daily life now and, by the way, an important component of the whole edition process: the addition and integration of pictures into the final document, when our book or article in question contains images, and not just mathematical figures. Those image files could thus be added very easily to the LATEX document in the process of compilation of the corresponding file.

Anyway, the enormous possibilities, the final evolution of the brilliant idea of first encoding an article or book in a LATEX file, then transmitting this file from computer to computer through the internet, and finally reproducing the whole book or article by compiling the file important to remark, without ever losing not the least single bit in the whole process!—had such an incredible potential that their uncountable applications could not be foreseen in the 1990s.

And the question is now, why was it so? What on Earth makes this concept, this very simple idea so extremely powerful and endlessly fruitful?

7 The Uncountable Present Uses of Cyberspace

The key hint for an answer in two sentences. The initial solution found to the problem of preprint distribution was to reduce the content of the preprint to a code, without any information loss. The written pages, possibly including mathematical plots ('figures'), that is, all the information of the preprint (or book) pages was encoded in a file. After sending this file through the internet, the end computer got this file from which the whole content of the initial object was recovered. In fact, the initial object itself! or one completely indistinguishable from it, when the file was printed. And now here comes the revolutionary idea. Just looking around you, there are many different things that could be codified, at least in principle, e.g., their essence reduced to a (more or less long) file, which might be transmitted and then reproduced at a position at the other end of the World (given such a wide network would be available).

The first addition to the original issue was the incorporation of images to the text/plot files, in other words, image encoding. This was done by optical scanning, as already discussed. In this respect, the following very important consideration is here in order. The number of letters in any alphabet is always finite, even if you want to encode those of all languages written on Earth, and in all possible different fancy fonts that may have been invented (of course handwriting is a completely different story, see later), so that you can choose the one you like more as final result, for your purposes. All this gives, in the end, a finite number of possibilities, which can be easily encoded and stored in personal computers of an ever-growing capacity.

Function plots are already a bit different, since they rely on appropriate mathematical programs for the approximation of function plots (usually Mathematica, Maple, or the like), which should be involved in the play. They use their own techniques for approximating any given function to a desired accuracy; but this works again pretty well. Images, on the contrary, are a completely different affair.

How do you capture the essence of a picture? Certainly, this has been done many times, with more or less success and under different circumstances, by painters, and later with the help of photo cameras. Now the problem is to digitize this, to encode this fabulous amount of information under the form of a binary file, as in the original case of the preprint. The problem is of a truly profound mathematical nature: to convert a continuous many-dimensional variable into a discrete one, to codify using 1's and 0's a continuous quantity that has, in principle, infinitely many possibilities, as the infinitely many different colors of a Renoir painting at infinitely-many positions (dots) on a canvas. Now, this can be done, but the clear difference with the simple case of the information contained in an ordinary book or preprint (without pictures) is that with images the codification of its content is no more lossless. By approximating the infinitely many possibilities by a finite number of pixels you will always loose some information. A good codification procedure will have a minimal loss with an also minimal length (or 'weight') of the resulting code or file. Optimal compression is a very essential concept that comes here into play and which, given the importance of this development, has quickly developed into a flourishing industry worldwide (Computer Science Stack Exchange Forum 2012).

But let us go back to our initial description and try to recapitulate. We scientists started by trying to improve our new inventions of TEX and the web by including a picture or two of some simple device or illustration, appearing in our preprint or booklet, and sending both together, always with pure scientific purposes. But immediately afterwards normal people realized the enormous power of this new toy; they bought themselves a computer and went on to send through the internet family pictures, and then movies (why not? that are nothing else but collections of pictures), and so on and so forth. And finally the computer was unnecessary and they started to do this all the time from anywhere with their small and very fancy mobiles. The initial, triangular thin path connecting the computers in Los Alamos, CERN and DESY evolved into an ever growing network of ways, roads, and highways. To end up becoming a truly global network, like that of a gigantic spider covering the whole Earth, which was given the most appropriate name one could ever have thought of: The World Wide Web (www or just Web) (W3 2016).

But this is of course not the end, rather the mere beginning. As the next possibility after sending pictures and movies of all kind you will immediately think of sending three-dimensional objects. Any item will be reduced to small components, and with the advent of 3D printers, after recovering the code of the device at destination you can 3D print it and obtain an instant copy of your favorite toy or working machine anywhere. And after you connect this development with no less important ones taking place in all other fields of scientific, humanistic, cultural, artistic or technological knowledge, the power of this "transmission idea" becomes overwhelming, unstoppable: we miss the words to describe it.

At the level of remote control of a device, when you do not transmit a physical object, but only an order to a remote machine, this procedure is already extremely helpful. Think, just of medicine: medical operations, the most delicate surgeries, diagnostic and control processes in medical praxis are now routinely carried out from the distance by the best of the specialists available. I am sure the reader, on his/her own, will find lots and lots of different examples to add here.

8 The Dangers of Cyberspace

But not everything is positive, with the Cyberspace. The dangers of this new, ubiquitous phenomenon are quite obvious, and have been stressed very often, e.g., by the Wall Street Journal. I copy from there:

> Do we know what our kids are doing online? What do you think are the best ways to protect your kids in a world where the walls between online and online are coming down? (Ilgenfritz 2008).

Sure, on general grounds the situation is not new at all. It has been repeated in human history every time a new invention has appeared; starting from the very first remarkable discoveries, as those of the wheel (known already in the Chalcolithic, 4000 years BC, before the Early Bronze Age), and the lever (known to Archimedes,

3rd century BC: "Give me a place to stand, and I shall move the Earth with it", and continuing with much more recent crafts, as the airplane (flown for the first time by the Wright brothers in 1903), and a lot many others. In spite of the generically good purposes of the authors of all these inventions, always aimed at easing human work and improving performance and possibilities, soon all those new instruments were used by the armies of all ages to develop evermore powerful deadly weapons.

In the case of the Cyberspace some of the dangers are much subtler, and most of them have nothing to do with the original purposes of the Arpanet, the American military network at the very origin of the internet. It is certainly true that among the most fearful dangers of the Cyberspace are its possible uses by all sort of international terrorists (Weimann 2015), organized criminals (Broadhurst 2013), mafia cartels (Agarwal 2006), drug dealers (Ngo 2014), and so on. But there are other very serious dangers of the internet that, having also to do with crimes, those are of a very different kind, as cyberbullying and harassment, in its various forms (leading too often to suicide), sexual predation targeting children in chat rooms, too easily accessible pornography, damaging person's reputation (Kam 2007), and several other. All those issues have been sufficiently addressed, and in much depth, in the references just given above and elsewhere, also feasibly in other chapters of the present book. I will do not deal with them in any more detail.

Instead, I want here discuss, in a little bit more extent, some rather subtle point that, however, cannot escape anybody's attention. Everywhere, every day, while just walking, or traveling to work, or sitting on a bench in a park for a short rest, eating at a restaurant, summing up, anywhere, anytime, you will be surrounded by people of all ages and condition attentively looking at a small device, sometimes (but not that often) talking to it, while they hold it on one hand, namely a fancy cellphone. An increasingly large number of people heavily depend on this device: their personal connection to the Cyberspace. Some of them have become absolutely fascinated by it, they have even got addict to it, and cannot stop looking at their cellular screen constantly. In a word, we could say that these people actually live in the Cyber world, rather than in the real one.

We may certainly view, at least some of them, as actual slaves of the Cyberspace: a new, contemporary form of slavery (Blume 2015) where individuals are not bound any more by ropes or heavy chains, against their will, but where they freely choose to be bound to this web of very thin, invisible, but extremely strong threads (for all we may judge): the internet. I will not go into the problem in much further detail, in the first place because it is none of my competences. But I did want to rise the point because it is a very serious one that really worries me and cannot go unnoticed. As I said before, it was no surprise, on the contrary, it actually was to be expected, from previous experiences, that the military on one side and the criminal forces, on the other, would immediately try to make use of the new invention at discussion here, as it has always been the case in past history. But this other issue I am addressing now is a brand new development, a kind of new illness without any precedent with past inventions. Possibly it will just be a strong but passing fever but this is not all that clear right now.

Of course one might argue that, in principle, there is nothing wrong with making your free choice to join the Cyberspace. Everybody can make extraordinarily good use of the immense possibilities it offers, as conveniently stressed in the preceding sections and many people do so, in fact, even a certain percentage of those internet addicts, I would dare say. But what is really worrisome is the huge amount of people, young and old, who, quite on the contrary, are too much distracted by the internet with very trivial (if not dangerous, as seen above) issues, which rob them of all their precious time. So that they are left with no single minute to carry out any positive task: learn, work, concentrate on some important issue, etc. Maybe this formulation is taking things to the extreme, but anyhow it is not so far from truth, in regretfully too many cases.

9 The Future of Cyberspace

I do not dare to make a list, not even a sketch of the extraordinarily huge number of future capabilities of the internet concept (Breene 2016). The advances in genetics and medicine, with the discovery of the genetic code of animals and humans (Radford 2003), the creation of human organs from cells (now not necessarily mother cells) (Weisberger 2016), the physical and technological exploration of the nano, pico and femto scales (Bradley 1997), … All these investigations rely on mathematical modeling, on the reduction of a real object or device to a certain file, which encodes all relevant information of the original (sometimes absolutely all of it). And once you are there, you can then send the file anywhere on our Planet or even on tiny little ships, the size of a postcard or less, to visit different stars and extraterrestrial planets, in a universal journey soon to be undertaken by the human society (Feltman 2016).

With the use of quantum technology, the possibilities to encrypt information (Preskill 2016) in a secure way and those of teletransporting it to far distances could make it possible, in a not so far away future, to send egg cells or nanocomponents, or pico-machines and devices (still to be conceived), with the aim to colonize our Universe. And this would happen in a very different way to the one imagined until very recently, which always involved big, manned, shining spaceships. However, one should be very careful with the quantum world (Merali 2015). In fact, in this case the result is a true teletransportation, since the very important no-cloning theorem prevents the cloning by the Einstein-Podolsky-Rosen procedure (Fine 2013) of any quantum state. In fact, the brand new state emerges, far away, only and at the very same instant when the original state disappears, ceases to exist.

The concepts of artificial intelligence and of artificial life will be further exploited, with consequences that are difficult to assess in a rigorous way yet. Should we be afraid of robots? And by this I mean, of replicants, of perfectly working copies of a human being, capable to pass all intelligence tests (Veselov 2014) available; what will make them indistinguishable from a real person. This has already appeared, and quite often, in science fiction movies. But what I mean here is that, at some point, this issue will no more be fiction science, but true, well established science.

Will we humans be able to live forever (or at least for a very, very long time) in one of these new bodies?—our self-having been captured, as a whole, in a sort of extended code and then having been put inside the replica, right here or millions of miles away. Will our soul be eventually replicated, also? And that, with all our memories, and with all of our feelings, as love, and fear, and freedom, and anger, too?

10 Conclusion

We started our project by aiming at the codification of a mere set of letters, of words and sentences—of more or less important scientific results—and at sending them 'for free' to distant places, and have now ended up by pretending to codify our own body and soul. So powerful is the Cyberspace concept that resulted as the end product from the process. An extremely simple, but enormously clever, solution to a very down-to-earth problem, just aimed at saving time and money, was the tiny seed that grew up unsteadily, to become a monster network of ever growing, immense possibilities.

Acknowledgement The author was partially supported by MINECO (Spain), Project FIS2013-44881-P, by CSIC, I-LINK1019 Project, and by the CPAN Consolider Ingenio Project.

References

Agarwal S (2006) Mafia invades cyberspace. Business Standard. Retrieved from http://www.business-standard.com/article/technology/mafia-invadescyberspace-106080101122_1.html
AvHumboldtE (2016) Asociación Alexander von Humboldt de España webpage. Retrieved from http://www.avhe.es/
AvHumboldtF (2016) Alexander von Humboldt Stiftung/Foundation webpage. Retrieved from https://www.humboldt-foundation.de/web/start.html
Berners-Lee T (2016) Personal webpage. Retrieved from https://www.w3.org/People/Berners-Lee/
Blume B (2015) Why the internet is a modern form of slavery. Retrieved from https://antsle.com/2015/11/04/why-the-internet-is-a-modern-form-of-slavery/
Bordag M, Elizalde E, Kirsten K (1996) J Math Phys 37:895 [hep-th/9503023]
Bradley D (1997) Yotta nano pico femto atto zepto yocto. Retrieved from https://www.sciencebase.com/science-blog/yotta-nano-pico-femto-atto-zepto-yocto.html
Breene K (2016) What is the future of the internet? Retrieved from https://www.weforum.org/agenda/2016/01/what-is-the-future-of-the-internet/
British Council (2014) 80 moments that shaped the world. Retrieved from https://www.britishcouncil.org/80moments/
Broadhurst R, Grabosky P, Alazab M, Bouhours B, Chon S, Da C (2013) Crime in cyberspace: offenders and the role of organized crime groups. Retrieved from https://papers.ssrn.com/sol3/papers.cfm?abstract_id=2211842

ClayPoinc (2016a) Millennium problems, Perelman's solution. Clay Mathematics Institute, Retrieved from http://www.claymath.org/millennium-problems-poincaré-conjecture/perelmans-solution

ClayPoinc (2016b) Millennium problems, Poincaré conjecture. Clay mathematics institute, Retrieved from http://www.claymath.org/millennium-problems/poincaré-conjecture

Cognola G, Elizalde E, Nojiri S, Odintsov SD, Sebastiani L, Zerbini S (2008) Phys Rev D 77:046009 [arXiv:0712.4017 [hep-th]]

Computer Science Stack Exchange Forum (2012) Computer science questions: is there any theoretically proven optimal compression algorithm? Retrieved from http://cs.stackexchange.com/questions/3316/is-there-any-theoretically-proven-optimal-compression-algorithm

Desy (2016) View of the DESY auditorium, with the piano on the right corner. Retrieved from http://it.desy.de/e5/e145041/e145129/e145151/e149599/index_eng.html?preview=preview

Elizalde E (2012) Ten physical applications of spectral zeta functions. In: Lecture notes in physics, 2nd edn, vol 855. Springer, Berlin

Elizalde E (2016) Personal webpage. Retrieved from http://www.ice.csic.es/personal/elizalde/eli/eli.htm

Elizalde E, Odintsov SD, Romeo A, Bytsenko AA, Zerbini S (1994) Zeta regularization techniques with applications. World Scientific, Singapore

Elizalde E, Nojiri S, Odintsov SD (2004) Phys Rev D 70:043539 [arXiv: hep-th/0405034]

Elizalde E, Nojiri S, Odintsov SD, Ogushi S (2003) Phys Rev D 67:063515 [hep-th/0209242]

Fax authority (2016) webpage. Retrieved from https://faxauthority.com/fax-history/

Feltman R (2016) Stephen Hawking wants to use lasers to propel a tiny spaceship to Alpha Centauri. Retrieved from https://www.washingtonpost.com/news/speaking-of-science/wp/2016/04/12/stephen-hawking-wants-to-use-lasers-to-propel-a-tiny-spaceship-to-alpha-centauri/

Fine A (2013) The Einstein-Podolsky-Rosen argument in quantum theory. Retrieved from http://plato.stanford.edu/entries/qt-epr/

Ginsparg P (2016) Personal webpage. Retrieved from http://infosci.cornell.edu/faculty/paul-ginsparg

Ilgenfritz S (2008) The real dangers of cyberspace. Retrieved from http://blogs.wsj.com/juggle/2008/08/22/bullies-and-instant-messagethe-real-dangers-of-cyberspace/

Inspire HEP (2014) Top cited articles of all time (2014 edition). Retrieved from http://inspirehep.net/info/hep/stats/topcites/2014/alltime.html

Kam K (2007) 4 dangers of the internet. Retrieved from http://www.webmd.com/parenting/features/4-dangers-internet#5

Knight D (2014) Personal computer history: the first 25 years. Retrieved from http://lowendmac.com/2014/personal-computer-history-the-first-25-years

Knuth D (1969) The art of computer. Addison-Wesley Pub. Co. Retrieved from http://www-cs-faculty.stanford.edu/knuth/brochure.pdf

Lamport L (1986) LATEX: a document preparation system. Addison-Wesley, Boston

Larivière V, Sugimoto CR, Macaluso B, Milojević S, Cronin B, Thelwall M (2013) ArXiv e-prints and the journal of record: an analysis of roles and relationships. Retrieved from https://arxiv.org/ftp/arxiv/papers/1306/1306.3261.pdf

Lewin W (2016) MIT lectures. Retrieved from https://www.youtube.com/playlist?list=PLyQSN7X0ro203puVhQsmCj9qhlFQ-As8e

Merali Z (2015) Quantum physics: what is really real? Retrieved from http://www.nature.com/news/quantum-physics-what-is-really-real-1.17585

Milleniump (2016) Millennium Problems. Clay Mathematics Institute, Retrieved from http://www.claymath.org/millennium-problems

Ngo C (2014) Illicit drug deals multiply on the dark net (Ipsnews, 2014). http://www.ipsnews.net/2014/07/illicit-drug-deals-multiply-on-the-dark-net/

Penrose R (2016) At Starmus 2016. Retrieved from https://www.plos.org/open-access/ http://www.starmus.com/dt_team/roger-penrose/

Perelman G (2002) The entropy formula for the Ricci flow and its geometric applications. arXiv: math/0211159 [math.DG]

Perelman G (2003a) Ricci flow with surgery on three-manifolds. arXiv: math/0303109 [math.DG]
Perelman G (2003b) Finite extinction time for the solutions to the Ricci flow on certain three-manifolds. arXiv: math/0307245 [math.DG]
Perelman G (2014) Documentary about the reclusive Russian mathematician who proved the Poincare conjecture and then refused a million dollar. Retrieved from https://www.reddit.com/r/Documentaries/comments/26jcb1/grigori_perelman_cc_2014_documentary_about_the/
Powell K (2016) Does it take too long to publish research? Retrieved from http://www.nature.com/news/does-it-take-too-long-to-publishresearch-1.19320
Preskill J (2016) Quantum information theory. Retrieved from http://www.theory.caltech.edu/people/preskill/ph229/notes/chap5.pdf
Radford T (2003) Human code fully cracked. Retrieved from https://www.theguardian.com/science/2003/apr/14/genetics.research
Smith T, Flückiger F (2016) Licensing the web. Retrieved from https://home.cern/topics/birth-web/licensing-web
Southwick F (2012) All's not fair in science and publishing. Retrieved from http://www.the-scientist.com/?articles.view/articleNo/32287/title/Alls-Not-Fair-in-Science-and-Publishing/
Starmus (2016) Starmus festival 2016. Retrieved from http://www.starmus.com/the-starmus-festival-2016-presents-the-programme-for-its-third-edition-beyond-the-horizon-a-tribute-to-stephen-hawking/
Veselov V (2014) Computer AI passes Turing test in 'world first'. Retrieved from http://www.bbc.com/news/technology-27762088
Weimann G (2015) Terrorism in cyberspace. Woodrow Wilson Center Press, Columbia University
Weisberger M (2016) Body parts grown in the lab. Retrieved from http://www.livescience.com/53470-11-lab-grown-body-parts.html
WikiArxiv (2016) Arxiv. Retrieved from https://en.wikipedia.org/wiki/ArXiv
W3 (2016) The World Wide Web webpage. Retrieved from https://www.w3.org/

Author Biography

Prof. Dr. Emilio Elizalde was born in Balaguer (Spain) in 1950. He obtained his MS in Physics, MS in Mathematics, and Ph.D. in Physics from Barcelona University, first and last with Extraordinary Master and Doctorate Awards. Elizalde was Humboldt Fellow in Hamburg and Berlin (Germany), and SEP Fellow in Japan. He held visiting Scholar Appointments and/or Research Contracts at Hamburg U, the MIT, the CfA at Harvard U, PennState U, Trondheim NTNU, the KEK, Hiroshima U, Jena U, Leipzig U, the NTZ, and Trento U. Professor of Mathematics and Physics at Barcelona University for over twenty years. Since 1993 he is a member of the National Higher Research Council of Spain (CSIC), where he is a Senior Research Professor and Principal Investigator of various projects—including a top rated Consolider one (group project)—and executive board member of both a Consolider-Ingenio 2010 (CPAN) and a European Union project (CASIMIR). Elizalde is the founding leader of the "Theoretical Physics and Cosmology" Group of ICE-CSIC and IEEC, highly recognized internationally, unchallenged in Spain in its field, and leading the ranking in Normalized Impact Factor of SCIMAGO World Reports 2009–13. He got an Honorary Professorship from Tomsk TSPU University, Russia, and is recipient of the Gold Medal of TSPU. Elizalde has published groundbreaking works on zeta functions, the Chowla-Selberg formula, and cosmology. One zeta function is named after him. He is very proud of his former students, a number of them being highly reputed scientists now.

Narrative Mapping of Cyberspace. Context and Consequences

David Harries

Abstract Cyberspace is barely two decades old. Yet it is already globally pervasive, powerfully disrupting perceptions and realities in the legacy spaces; on the land, at sea and in the air where human beings live, move and work. The pace at which its influence is spreading and intensifying is amazing; the number and reach of the consequences arguably even more so, and they continue to emerge, mash-up and surprise. For humanity and its planet, an acceptable future depends on 'seeing' and understanding Cyberspace well enough to do two things; manage and exploit it successfully in the present, and make timely, flexible preparations for a future that is uncertain, except in that it will be different to today, in no small part because it will be substantially shaped by the state of and actions in the Cyberspace. This chapter is a first attempt to explain why Cyberspace has become so important so quickly and describe briefly the most meaningful of its initial consequences, all with the aim to promote strengthening the good in Cyberspace while keeping the bad in check.

Keywords Context · Consequences · Borderlessness · Vulnerability

1 Introduction

Mapping 'the Cyberspace' calls, arguably, for three distinct but related maps. Each one needs a legend or key. One map is of Cyberspace structures and processes. Put in the simplest terms this map answers the question; What is the Internet? A second map shows what is taking place—what is being done—*in* Cyberspace. The third

Submitted: 4.10.16; Accepted: 3.11.2016.

D. Harries (✉)
Chair Canadian Pugwash Movement, International Practitioner of Security Foresight, Kingston, Canada
e-mail: jdsharries@bell.net

J.M. Ramírez and L.A. García-Segura (eds.), *Cyberspace*, Advanced Sciences and Technologies for Security Applications, DOI 10.1007/978-3-319-54975-0_2

map is of the context in which Cyberspace exists, its activities take place, and their consequences play out. It is a 'motion' picture of trends and drivers of cyberspace's character and consequences in an increasingly connected, but uneven, world.

The legends for these maps are works in progress, for three main reasons. First, the Cyberspace is young; barely out of its teens. Social media, arguably the most globally influential element of Cyberspace, is even younger. Second, it exists in a world stressed by accelerating, often shocking, change. And, third, Cyberspace is a self-organizing intangible that defies proactive control.

All three legends need to cater for the scope, spread and significance of Internet activities, which are encouraged most strongly by the technology imperative (Buzan 1987). Any legend for a map of Cyberspace context calls for metrics and pictures that depict matters as grand as globalization and as granular as the well-being of individuals. For this reason, it was decided to focus, narratively, on the context and consequences of Cyberspace; what is obvious today and what can be foreseen for plausible futures.

This chapter attempts to explain how and why a 'space' that only recently came into existence has so quickly become so pervasive and so powerfully disrupted perceptions and realities for the legacy spaces; land, sea and air, of human existence; of living, moving and working. The explanation is incomplete, and will remain so until a number of questions can be answered in more detail than is possible today. Finding useful and appropriate answers, and identifying the important connections and overlaps among them, will be challenging given Cyberspace's ephemeral and ever-changing nature.

This fact leads to the author to conclude that more time and effort needs to be routinely devoted to Cyberspace Foresight. The Cyberspace became globally important very quickly, and the pace at which its influence is spreading and intensifying is amazing. For humanity and its planet, an acceptable future depends on 'seeing' and understanding Cyberspace well enough to do two things; manage and exploit it successfully in the present, and make timely, flexible preparations for a future that is uncertain, except in that it will be different to today, in no small part because it will be substantially shaped by the state of and actions in the Cyberspace.

2 Cyberspace Context

2.1 Context Background

In 1984 William Gibson, who coined the term 'cyberspace' on 1982, published his novel Neuromancer. It tells the story of a washed-up computer expert hired by a mysterious employer to pull off the ultimate hack. Neuromancer is the name of one of two Artificial Intelligence characters in the story, both of whom would be at home in 2016.

Until, arguably, the 1990s, if someone, say a whistle-blower, wanted to tell another person something 'privately', without others knowing about or seeing the message, the 'brown envelop' was standard practice. If the information needed to be sent to many people, copies had to be made, and individual envelopes addressed. Today, 'snail mail' whistle-blowing is rare,[1] not only because it is so slow, but because a message can now be sent to any number of others, in the Cyberspace, in less time than it takes to manually address one envelope. In addition, the sender can choose from a variety of Cyberspace ways and means to remain anonymous.

The Cover Story of the June 2011 Consumer Reports magazine was titled 'Your Security'. The first three articles were: (1) 25 things cops and crooks say you have been doing wrong, (2) Why your accounts are vulnerable to thieves, and (3) Door locks; all conventional information and advice seen well before 2011 and little changed, except for updating technology and cost factors for 2011. The *fourth* and last article in the package was: 'Online Exposure: Social networks, mobile phones and scams can threaten your security'. The list of 'details', highly focused on 'abuses' on Facebook and careless mobile phone use, ended with the statement: "The persistence of Internet threats makes it important to use security software. In our tests we found that free anti-malware programs should provide adequate protection for many people." The next article in the magazine was a seven-page Report: Portable Computers The new choice: tablet, laptop, or netbook. Not once does the word 'security' appear.

The Special Report of the July 12th 2014 edition of The Economist, was titled Cyber-Security: Defending the digital frontier, begins with a reference to William Gibson's coining of 'cyberspace'. It highlights the paradox that, on the one hand the internet has "...brought tremendous benefits to everybody who uses the web..." and on the other that "there is a darker side this extraordinary invention", in terms of all of commerce, national security, public safety, conflict and recourse for abuse. The 14 page narrative is virtually all about the 'darker side' of the Internet, and on people's willful blindness to its Janus-like nature; the threats and the opportunities of weak source codes, hacking, exploding, unconstrained and unpredictable connectivity, negative externalities, cyber-crime and cyber-extremism, and plausible deniability. Nothing in this 2014 Report needs to be changed for 2017 except, maybe, its closing message; Prevention is better than cure, which is outdated. 'Defending the digital frontier' can no longer be done only, or best, by prevention. As soldiers know, and as the growing community of commercial and state-sponsored organizations have taken to heart, offence is the best defence.

Today, it is a rare for newspaper or newsmagazine not to have something on the status, security or activities in Cyberspace, so embedded, so 'normal', has it become in everyday life, world-wide.

[1]Rare, but not never. Parts of Donald Trump's 1995 tax returns were mailed in August 2016 to a NYT reporter. See http://www.nytimes.com/2016/10/02/us/politics/donald-trump-taxes.html?hp&action=click&pgtype=Homepage&clickSource=story-heading&module=b-lede-package-region&region=top-news&WT.nav=top-news&_r=0.

2.2 Contemporary Context

The arrival of the Cyberspace offers unprecedented power to individuals and organization that, beforehand, could neither attain nor exercise it. Today, everyone with an internet connection has significant power. The most powerful are those who first place 'information' in Cyberspace. If the information is 'true', those who first exploit it in productive ways can become a more powerful competitor in their field. Information that is not 'true' can both engender power, and weaken it. Its receipt weakens the relative power of those who do not know it is false information, the more so if they use it in the belief that it is true. Recipients who know the information is false have a choice. On the one hand, they may see it as a threat to their reputation or resources that demands the time and effort (power) to immediately rebut or correct it. On the other hand, if the untrue information is not a threat to one's current well-being and intentions, the falsehood may represent nascent power that can be used in the future, possibly against its sender.

There are many choices for the name of the 'age' ushered in by the arrival of the Cyberspace: the digital age, the information age, the internet age, the computer age, the connected age, the Google age, the emoji age, the cloud age, the smartphone age, the data age, the Facebook age, the robot age, the post human age (Aeon 2016). The more names it is given, the more vaporous it seems, but whatever age has arrived, it is definitely one of information and data glut. The glut is growing exponentially as more people and organizations use systems and servers that are becoming faster, and smarter. It has become possible to find in Cyberspace masses of material on virtually any subject, in tiny fractions of a second.

However, reading and filtering all that can be found, to identify the parts that can be used or that need defending against grows ever more humanly challenging. Evidence of this fact is a recent Aeon report:

> Since 2010 Twitter has been providing the library with every tweet that's been posted publicly. It was supposed to be an archive for researchers. Managing that massive data dump, however, has proven to be a challenge the library has no idea how to handle it (McGill 2016).

The situation is a prime driver of development and deployment of faster and smarter Cyberspace tools. The number of individuals and organizations both active in Cyberspace and keenly aware how empowering their engagement can be, also is rising exponentially. The rise is self-reinforcing/self-sustaining. It is human nature to not want to be behind, or to be left out, or to risk being ignorant, and especially to be so described. Cyberspace has spawned a new metric for the old mantra 'keeping up with the Jones'.

Virtually everywhere on earth is under stress. There may be a few humans who are calm and have valid reasons to be so, but the great majority are, depending on the time and their circumstances, one or more of concerned, challenged, fearful, threatened, under attack, flailing, failing, defeated, or victorious (for reasons both real and perceived). The variety of the causes, degrees, validity, longevity and durability of stress, and of the response to it, has left the human condition massively uneven globally, arguably more uneven than at any time in human history and becoming ever more so. The manifestations of the unevenness, whether in terms of freedom, equality, stability or security are, thanks greatly to the Cyberspace, 'open source' information in real-time.

Globalization has effectively ended the 'sovereignty' of the infrastructure of each nation. The merit, value and integrity of infrastructure in one state, even the built parts, are no longer best measured primarily by and in that state, but in terms of the costs and the benefits relative to and shared with other states. Even the most powerful and richest states can no longer claim to be truly self-sufficient. Global supply, distribution and transportation chains, global wicked problems such as climate change, global monies (bitcoin) and money-moving (blockchains), and the global Cyberspace in which all are engaged and enabled, have relegated national self-sufficiency to the status of at best an occasional, temporary strength, and at worst an historical concept still governing national policy.

Extremism is a global phenomenon that appears in many forms and plays out in many fields. Cyberspace is where virtually all extremists market, demonstrate, contest and preach. One extremist's statement or action can be globally known in seconds. Since there is no recall or removal from (somewhere in) Cyberspace, the message or picture becomes one of an ever increasing mountain—glut—of eternal facts. In technology's distant past, it was sometimes possible to erase or silence news of a fact or event that those with necessary power and authority to do so did not want to become public knowledge. Today, even the most undemocratic countries, ones that go to extremes to stop their people from knowing about others, or the outside, can no longer totally seal their borders. The borders that Cyberspace recognizes are different to and usually at odds with the borders on the maps used by states and all other actors for whom 'international' retains a tangible meaning.

Long ago, crime and punishment were local, clearly defined and directly related. Over time, laws, the authority to monitor and enforce them, and justice, became more complicated and complex. Today they are ambiguous and incomplete, in large part because they have failed to keep pace with globalization dynamics. No law exists today that applies world-wide, or to all of any of the many recognized regions, or to the geopolitical stew of states, para-states, non-states and pseudo-states. Public organizations, nonpublic organizations, private individuals and corporate actors are all different in legal terms, even if in the same state. In addition, laws that apply to some of them may conflict with some that apply to others. There is no global body with the mandate to harmonize even the most costly and damaging of those contradictions. 'Law' is still made very much as it was in the 19th century, but governments from the very best to the very worst are all exploiting Cyberspace to promote their interests of the day.

Unsurprisingly, dictatorships are more effective at making laws, dealing with crime and meting out punishment in these times of context compression, than are democracies. The latter, even those using the internet in enlightened ways, are slowed and obstructed by democratic standards and practices from fully exploiting Cyberspace characteristics. This inherent power-lessness is intensified by the 'information glut' referred to earlier. The ever-increasing mass of necessarily relevant information deserves 'fair' and full—i.e., democratic, review by all stakeholders before a decision is made about a law or a punishment. Widening gaps in relevance and content among laws, regulations and standards are further delay legal process and judgement, leaving democratic governments ever more stressed by the tensions between fairness and justice.

The small world—that Cyberspace and technology have co-provoked—is growing more and more crowded, not only with people but with barriers. More 'silos', filters and barriers—physical and virtual—are being erected or re-erected. In part this is a reflection of the unhappy state of the world's geopolitics and of security fragmentation. Each silo, filter or barrier takes up space that cannot then be a 'commons' where people can live, move or work freely and productively. The increase in barriers is due to one or more of misunderstandings, inappropriate biases and assumptions, selfish interests, real and perceived fears of weaknesses in national security and public safety, context incompetence, and survival uncertainty. The barriers have many forms, among them; physical walls, trade protectionism, censorship, surveillance, ethnicity, education, and religion. The more barriers that go up, the more effort and resources that are deployed to monitor and defend them, and the less 'openness', equality and opportunities there are for more and more people. The inevitable result; and the more so as activities in Cyberspace highlight the trend, is greater likelihood of significant and substantive differences among individuals and peoples as space to live, move and work free from fear and want shrinks.

On land, crowding is being provoked by an even more pervasive trend; the continuing increase in the number of sovereign states, each of which insists on borders on maps, their side of which they control. New borders reduce the space for 'commons'. The global mantra of entitlement to self-determination and independence reinforces the growth in physical borders as several to many more of the thousands of nations aspiring to statehood feel justified in doing so. How many of them will succeed in joining the 193/200 existing states, and when in relation to population growth now estimated to be heading to 11 billion, or more.

3 Cyberspace Consequences—General

The 'arrival' of cyberspace has already meant lives the majority of human beings has a 'family' of new, changed and changing challenges—both threats to be faced and opportunities to be seized—in the traditional spaces where everyone lives, moves and works. How bad are the threats and how good are the opportunities will determined by an unpredictable, and unmeasurable combination of our proaction for and our reaction to them.

Cyberspace needs regulation. Considerations of who, how, and at what cost management, regulation and control can be established have barely begun. But, until they are effectively in place, individually and collectively, identifying, deploying and governing the effort to deal with the planet's existing, and arguably intensifying 'wicked problems'; among them Climate Change, Conflict and Context Compression may be more confusing that constructive.

Any map of consequences of the CS and its activities will show a living, self-organizing mash-up of what first and early on were seen as almost universally positive and bright impacts, and the increasingly more numerous less positive and darker pictures as its activities reached ever more deeply into ever more aspects of human life. The whole map is blurry, with little in focus even in the short periods between events and circumstance changes that redraw, recolor and reorient its features. Grey spaces on the map signal that there is still more to be learned about consequences provoked in the short Cyberspace past. Dark spaces unknown unknowns, and will be where the inevitable wild cards and black swans substantially and suddenly—in an artistic explosion—disrupt all or most of the map because not enough was known to put in place the ways and means to soften the blows.

4 Overarching Consequences of Cyberspace

In these early years of the Cyberspace, in a world where the *status quo* is shocked and disrupted with alarming regularity, and given that the writer is not a cyberspace expert, it borders on the presumptuous to claim that the following are, already, overarching, or permanent, consequences. However, even if they prove less august with time, they *will* feature on any map of the Cyberspace.

1. First, and foremost, anyone who is connected to the WWW, or whose lives depend on goods or services connected to it, is potentially vulnerable to any and all of the threats from the dark side of the internet. The more frequent the connection, and the 'smarter' the goods and services and the more connections among them, the greater the vulnerability.
2. What the Cyberspace provides; universal and enduring visibility of all in it, provides people with ever more 'knowledge'—true and false—about their concerns and fears and the challenges and threats facing them. This intensifies their already high levels of stress. Rising stress is promoting greater uneveness of the human condition globally, and reinforcing the trend—also apparently global—of rising numbers with mental health problems, primary among them being PTSD and abject desperation. The former was formally recognized only a few years before the Internet arrived, primarily as a soldier's mental health issue. Today, thanks to the Internet's ability to show everyone the most horrific things in near real time, and medical progress, virtually everyone has become a 'first responder'—if only mentally—to the evils of our times. Abject desperation is driving more people to risk all to improve prospects of survival with freedom, and others to see no alternative but to buy into extremists' ideologies, usually imposing conditions that are anything but free, and calling for behavior that precludes survival.

As Nicholas Carr writes "Technology promised to set us free. Instead it has trained us to withdraw from the world into distraction and dependency." (Carr 2016; Weintraub 2016).

3. 'The truth, the whole truth and nothing but the truth'. Long the standard for participation in processes to assess right and wrong and judge on the findings, this call has, in whole, become an impossible task. Veracity has been recontextualized to be a reflection of only what is known and being considered at a given place and moment in time. It has become a vision in search for more and better truth. The age of truthiness has arrived, an imperfect age both governed and driven by the only contemporary 'truth'; uncertainty.
4. The WWW both draws people together and drives them apart. Its contents can shift the relationship from positive to negative in the time it takes to read the latest email. Scanning the web, depending on one's reason (Harries 2016) for doing so, quickly demonstrates that there is more than enough evidence, opinion or speculation in Cyberspace to allow even the most rational of scanners to come off the fence on the side of either camp arguing about almost any issue.

5 Specific Consequences of Cyberspace: A Selection of Four

The consequences of the Cyberspace will continue to grow in number and therefore in their influence on everyday life. The positive 'good' ones are much discussed and generally agreed. However, negative 'bad' consequences receive less than their due, including accounting for the fact that many and arguably most Cyberspace consequences should be seen as offering a set of good and bad ones. The main problem in differentiating the two can be expressed by a paraphrase of a familiar statement: One man's terrorist (bad consequence) is another man's freedom fighter (good consequence).

> Mr. Snowden has seen significant success in his quest to start a public conversation about government surveillance. In 2015, the N.S.A.'s bulk collection of Americans' phone records, one of the programs he exposed, was ruled illegal and transformed by Congress. He has also won widespread support abroad, including from the European Parliament, which adopted a nonbinding resolution in October to protect him from prosecution and recognize him as a "whistle-blower and international human rights defender" (NYT 2016a).

This problem is exaggerated by all four of the overarching consequences briefly described above: Vulnerability, Stress, Truthiness, and IFF (identifying friend or foe), in large part because all are 'in motion'; sensitive to time and circumstance and inherent opaqueness.

Five consequence fields have been selected for brief, and therefore far from deserving, attention; Power, Conflict, Personal Well-being, Business, and Foresight. There are myriad, dynamic connections, overlaps, and influences among them, but doing that fact justice would demand a map with detail and fineness that is beyond the scope of this paper and the ability of the author. The fifth field; Foresight, is chosen not only because it one the author knows and practices, but to introduce the concluding section of this think-piece; a set of Questions whose full or partial answers will increase the likelihood that consequences of the Cyberspace now, those emerging, and those shaping and in the future, can be identified and understood in time to be appropriately managed.

The focus is on the 'bad', or at best the less good consequences, as these are invariably the ones demanding the most attention, soonest.

5.1 Power Consequences

Information is power? Information is power if it is used. For the first time in history a single individual can hold a nation or group of nations hostage to an uncertainty, or threaten or attack them in both tangible and intangible ways. Even a nuclear threat or an attack which—other than in the tragi-comical rants of the North Korean dictator—have not featured in internet exchanges, calls for the deliberate participation of several persons and substantial resources.

But it remains impossible to know with absolute certainty if an event in Cyberspace is the act of an individual, or of an organization, or of a state, or of a group of states, unless the actor confesses. Organizations and states rarely confess, even when caught. Individuals do; the most famous being Manning, Assange and Snowden, some even when not caught. In any case, plausible deniability exercised smartly offers any attacker more 'power' to deploy uncertainty.

Who or what is Guccifer 2.0? Who or what is? are? the Illuminati? Uncertainty has been enriched by the wholesale writing of history globally, an activity given wings by the Cyberspace where everyone can become an historian, not only as before the internet, only the victors or the rich and powerful with the wherewithal and discretionary time to put pen to paper and attract publishers. The many new histories have severely dented the reputation and therefore the power of all who, in the present, rose to power on the back of traditional, but incomplete or half-truths or outright falsehoods. The power of the new historians and of the communities their news spawns is already significant enough to counteract traditional state-based strengths (NYT 2016b).

The strong are now less powerful if they do not exploit the Cyberspace. Putin, the forceful leader of a huge country that is actively deploying cyber weapons, no doubt in part to make up for the host of problems; economic, social, political and military that are weakening Russia (Tsygankov 2016).

"Using both conventional media and covert channels, the Kremlin relies on disinformation to create doubt, fear and discord in Europe and the United States (MacFarquhar 2016)".

But exploiting the Cyberspace has its downsides, and raised many unprecedented questions that do not have durable answers. Even if officials of the US—far and away the most militarily powerful and globally deployed nation in history—obtain proof of who or what has used a Cyber tool or weapon to attack the country, or lied to them, or stolen from them:

> they may not be able to make their evidence public without tipping off Russia, or its proxies in cyberspace, about how deeply the National Security Agency has penetrated that country's networks. And designing a response that will send a clear message, without prompting escalation or undermining efforts to work with Russia in places like Syria, where Russia is simultaneously an adversary and a partner, is even harder (NYT 2016c).

Again, on Guccifer 2.0 and the Illuminati? Who are their friends. Who are their enemies? Are the two always enemies?

5.2 Consequences for Conflict

Booby traps have been a weapon for centuries. The arrival of the Cyberspace allowed them to move on from what were historically manually produced, in-place, crudely timed and simply controlled (if at all), victim-activated weapons for defence and protection or psychological effect to become a far more fearsome weapon controlled by the attackers. Since 2001, more NATO deaths, casualties and PTSD have been attributed to the now infamous IED (Improvised Explosive Device), often wirelessly controlled (RCIED), than to any other Taliban or warlord weapon (or to friendly fire accidents).

There are no principles of war for Cyberwar: "Mr. Obama often says the world of cyberconflict is still "the Wild West." There are no treaties, no international laws, just a patchwork set of emerging "norms" of what constitutes acceptable behavior (NYT 2016c)".

There are no norms for Cyberwar. Who admits what? Who is qualified (has the authority) and competent (the technical knowledge and skills) to speak? There are neither norms for evaluating cyber education and training, nor certifying those educated and trained. Does that mean anyone can be an expert?

Decisions on Cyberwar options are more guesswork than judgement. The situation is a duel of uncertain facts in a fog of uncertain second and third order consequences of unpredictable costs and benefits.

> In the Democratic National Committee* case, two senior (US) administration officials spoke on the condition of anonymity to discuss the options, ranging from counter cyberattacks on the F.S.B. and the G.R.U., two competing Russian spy agencies at the center of the current hacking, to economic, travel and other sanctions aimed at suspected perpetrators (NYT 2016d).

Way and means to exploit the Cyberspace to attack others; to disrupt, to destabilize, to destroy, to confuse, to discourage seem unlimited:

> every new case (attack) brings a new and imaginative way to weaponize cyberpower. Until November 2014, when North Korea hacked into the computers at Sony Pictures Entertainment in retaliation for a comedy that portrayed a C.I.A. plot to assassinate Kim

> Jong-un, the country's leader, no one seriously considered a movie studio to be "critical infrastructure." (NYT 2016d).

The United States is, by far, the world's most aggressive nation when it comes to cyberspying and cyberwarfare. The National Security Agency has been eavesdropping on foreign cities, politicians, elections and entire countries since it first turned on its receivers in 1952. Just as other countries, including Russia, attempt to do to the United States. What is new is a country leaking the intercepts back to the public of the target nation through a middleperson. Unlike the Defense Department's Pentagon, the headquarters of the cyberspies fills an entire secret city. Located in Fort Meade, Maryland, halfway between Washington and Baltimore, Maryland, NSA's headquarters consists of scores of heavily guarded buildings. The site even boasts its own police force and post office.

And it is about to grow considerably bigger, now that the NSA cyberspies have merged with the cyberwarriors of U.S. Cyber Command, which controls its own Cyber Army, Cyber Navy, Cyber Air Force and Cyber Marine Corps, all armed with state-of-the-art cyberweapons. In charge of it all is a four-star admiral, Michael S. Rogers.

> "Cyber Command itself has always been branded in a sort of misleading way from its very inception, Snowden told me. It's an attack agency. ... It's all about computer-network attack and computer-network exploitation at Cyber Command" (Snowden 2014).

The idea is to turn the Internet from a worldwide web of information into a global battlefield for war. "The next major conflict will start in cyberspace," says one of the secret NSA documents. One key phrase within Cyber Command documents is "Information Dominance (NYT 2016d)".

NSA was, fairly certainly hacked in recent times. Officials and a 'former TAO operator' stridently responded to the news, that included an offer to auction more NSA data than had already been exposed, with emotion: 'the auction is a joke' and that the auctioneer 'doesn't have everything' (Nakashima 2016). True? False? How do they know?

War is becoming more and more automated (La Pointe and Levin 2016). Gone are the days of 'line of sight' battle, front lines, and soldiers fighting other soldiers. Indeed, it is only two decades since communications technology created the conditions for the appearance of strategic corporals and tactical generals, so visible had the whole battlefield become; visible, because of the deployment of drones. Then came drones with weapons, then bigger drones with more weapons and longer ranges and staying time. Generals in Florida and drone pilots in the center of the US fought the battle in Afghanistan. In 2012, the USAF trained more UAV pilots than ordinary jet fighter pilots for the first time. As observable data becomes even more granular and algorithms improve, autonomous systems will be able to support commander decisions.

In contrast to the famous Hellfire equipped Predator, which is remotely piloted via satellites, the Global Hawk operates virtually autonomously, and is leading the shift to the next level of war at a distance. Advances in AI are speeding the arrival of all manner of autonomous armed robots. How autonomous these Cyberweapons can be (technology), will be (geopolitical decision), and should be (humanitarian and ethics concerns), are issues under intensifying scrutiny.

Asymmetrical warfare? ISIS in the Middle East is losing territory in Iraq and Syria, predominantly because US, and probably others' airstrikes are picking off its leadership and attacking important facilities and resources. ISIS therefore increasingly needs to "depend upon its "virtual planners"—members who operate in the dark spaces of the Internet—to inspire and coordinate attacks abroad" (Foreign Affairs 2016).

5.3 *Consequences for Personal Well-Being (Human Security)*

The Cyberspace has significantly complicated 'human security' and forced new metrics upon it. As willingness to acknowledge—see—the consequences of the Cyberspace and as understanding of them rises, the metrics are beginning to appear. Primary among them is a grudging rebalancing of the relative priorities of security and personal freedoms. There is a more realistic attitude toward electronic surveillance and its contested but unavoidable role in modern counterterrorism, and acceptance that 'liberty' will be less. This ever stronger reality particularly rankles Americans—remember the motto of the State of New Hampshire, on every vehicle license plate: "Live Free or Die"—but is making its mark much more widely.

> Again from Carr:
> Technology promised to set us free. Instead it has trained us to withdraw from the world into distraction and dependency. The culture that emerged on the network, and that now extends deep into our lives and psyches, is characterised by frenetic production and consumption—smartphones have made media machines of us all—but little real empowerment and even less reflectiveness. It's a culture of distraction and dependency (Carr 2016).

> The Swiss have seen the 'writing on the wall':
> The Swiss (recently) voted in favor of increased government surveillance. Fear of terror attacks trumped Switzerland's traditional wariness of government snooping. More than 65% of voters were in agreement with the law that gives the Federal Intelligence Service more power to tap phones, read emails, and use bugs and hidden cameras." (BBC 2016).

Privacy? Yahoo Says Hackers Stole Data on 500 Million Users in 2014 (Perlroth 2016). To what end? What should the millions of owners of the hacked accounts do? Who is accountable? (Satter and Cheslow 2016). "The spyware took advantage of weaknesses in Apple's mobile operating system to take complete control of iOS devices (…) YouTube may be teaching someone to spy on you (Kasulis 2016)".

Confidentiality? WADA systems have been hacked. Given the intense and continuing war against doping in sports, the ends for the hacker may seem obvious, but what are the thousands of athletes whose tests' data is no longer confidential to do? Has any of the hacked data been tampered with? Are whistleblowers safe?

> Confidential athlete medical data relating to last month's Rio Olympics has been hacked and published by a Russian cyber espionage group with the threat of more to come, the World Anti-Doping Agency (WADA) said. It identified the group as Tsar Team (APT28), also known as Fancy Bear. The www.fancybear.net website said it had information about a number of U.S. athletes, including tennis sisters Serena and Venus Williams as well as multiple gold medal-winning gymnast Simone Biles. WADA revealed last month that Russian whistleblower Yulia Stepanova's electronic account had been illegally accessed with a "perpetrator" obtaining details which would normally include her registered whereabouts (…) Personal safety and health? Stepanova, referred to in the previous paragraph, is in hiding in North America, having been forced to flee with her husband for fear of her life after helping reveal the biggest state-backed doping programme in Russia (Ruiz 2016).

Social Policy planning based on good data, reliably available?
"Australia has halted online collection of national census data after a website where citizens could upload information was subjected to repeated cyberattacks. The Australian Bureau of Statistics said its website had experienced four denial-of-service attacks, in which a torrent of automated requests is sent to overwhelm a site. The last attack, just after 7:30 p.m. on Tuesday, contributed to the overloading of a router, which led to the decision that night to close down online data gathering. The census, which occurs every five years, has been the subject of intense criticism and questions this year over whether the introduction of online data collection could leave Australians' personal information at risk (Ramzy 2016)".

Canada, a relatively safe and respected state, is increasingly challenged by Cyberspace to manage both global and national contexts in appropriate fashion (Bell 2016). Activists in Canada critical of Beijing have found themselves targets for intimidation. Notwithstanding the clarity of the situation, they apparently have no recourse—they are 'powerless'.

"Not long after Zang Xihong, 54, a prominent Chinese human-rights activist, emigrated to Canada 27 years ago, she said, she began receiving menacing phone calls from Chinese state security agents at her home in the Toronto suburbs. In recent years, she said, the harassment has grown more ominous. Her face and phone numbers have been digitally inserted into pornographic escort ads, she said; hackers have posted photos stolen from her computer; and articles have appeared online accusing her of embezzlement. She has also been sued by a man who claims she was responsible for his cousin's death in China. Zang said Canadian authorities had told her that they could take no action because most of those activities were protected free speech, leaving her powerless, she said, to escape the long arm of the Chinese government or its supporters (Waterloo Chronicle 2016)".

Buying and selling safely? "Another woman was also scammed by a man who matches pictures of Reid. Liat Feldman just moved back to Toronto after living overseas for 18 years. She says that she, too, was scammed out of $1100 at the same apartment on the same weekend as Langton. "You can't go onto the internet anymore and find legitimate apartments," she told CBC. "You don't know who you can believe" (CBC News 2016)".

On "The Current" a highly respected current events programme provided weekdays by Canada's national broadcaster, the CBC, the interviewee, who had just had his book published on the state of the global con game, remarked that the con business is exploding in Cyberspace, that there is little chance, and therefore fear of being caught, and therefore next to no fear of ever being punished (Tremonti 2016).

5.4 *Consequences for Business*

There are arguably three predominant consequences of Cyberspace for business; new business, new map and new players. A detailed analysis of the three and of others of importance is in preparation.

1. New business. The business of Cybersecurity business has changed. It was, in the beginning the somewhat esoteric activity of providing the goods and services to get on the internet and protect computers and their connectivity from viruses and malware and remove those that make it through, or that infected the internet resources of those who—for quite a number of years in large numbers—did not think it necessary to invest in protection. It then briefly moved to protection and identification of attackers. Quickly, with states or state-sponsored or authorized private organizations in the lead, it has achieved what can be termed war footing; active in overt and covert defensive and offensive and psychological operations with substantial geopolitical overtones.

   > For example, Mr. Obama has pressed President Xi Jinping of China to work with the United States and other nations to develop rules about the theft of intellectual property, and about not interfering with a nation's efforts to bring attacked systems back online. Attacking another nation's power grid in peacetime is considered out of bounds (Sanger 2016).

2. New Map. The 'map' of costs and benefits of Cyberspace has been completely redrawn, for 'the good guys' and 'the bad guys'. The good guys have been shown how costly are insufficient, out-of-date, unprotected internet systems. A few years ago the perceived threats and the costs of their being realized were little discussed, and little prepared for. It is no longer a matter of affordable fixes and embarrassment control,[2] but one of survival of the business, operationally

[2]In 2008, in a private discussion late one evening at the US Army War College in Carlisle, Penn with members of its Strategic Studies Institute, the author was asked if he had heard of the 'theft from the US Treasury'. He had not, and said so, upon which he was told, confidentially, that 'someone' had 'looted the US' of more than 1 trillion 'in the seconds it took for the defences to kick in'. The event was never reported in US media (or anywhere else for that matter, to the author's knowledge).

and or financially. Therefore, cybersecurity companies are very busy. And the bad guys who are having a field day. Global cybercrime costs are estimated as 400 billion in 2014, with a forecast of 2.1 trillion in 2019 (Morgan 2016). There is little doubt which side is 'winning'.

3. New Players. The number, scale, variety of goods and services on offer, and market for Cyber goods and services are all rising (Cybersecurity Ventures 2016). Construction, energy, mining, and transportation companies are enjoying good days supporting the building of Cyberspace infrastructure. Names such as Tara Global, Crowdstrike, and Kapersky Lab may not yet be 'family names' but everybody working in or who depends on their Cyberspace systems being up and running knows who they are. This includes the bad guys who are working hard to stay ahead of them, a mission made easier because they have none of the implied and explicit constraints of those who, to some degree, value democracy and the rule of law.

5.5 *Consequences for the Future*

It is impossible to state with confidence what the consequences of the Cyberspace are for the future. But, it is reasonable to suggest that consequences already seen and experienced and those that anyone with some imagination can see if they try signal more consequences that will be ever more complicated, connected, costly and valuable.

One is sure to be future installments from the Snowden files, Anonymous and Wikileaks, to mention only the three of the most famous sources.

> Assange said WikiLeaks plans to start publishing new material starting this week, but wouldn't specify the timing and subject. Speaking by video link to an anniversary news conference in Berlin, he said the leaks include "significant material" on war, arms, oil, internet giant Google, the U.S. election and mass surveillance. WikiLeaks hopes "to be publishing every week for the next 10 weeks" (Assange 2016).

Deduction: With every additional installment of masses of information from Cyberspace, the map of its consequences will need to be redrawn if it is to be of continuing use. The drawing of that map will intensify the competition among both those drawing its basic outline and those preparing the legend that will determine what the map shows … until the next change.

A second consequence is a requirement to think thoroughly about the future context, and not only of what is preferred or expected, but what is not, and what might be very bad. To start and to frame the process early on, the following questions should be asked, listed in no particular order:

1. Whither democracy?
2. Whither diplomacy, and the need to go abroad to 'lie for your country'?[3]
3. Whither statehood, and the role of government?
4. Whither Climate Change, and the response?

6 Conclusion

The Cyberspace has had, in two quick decades, enormous consequences for the planet and its inhabitants. There is no sign—none—that the power of Cyberspace's character and activities to provoke change; good and bad, will soon fade. Therefore, Cyberspace consequences for the planet and it inhabitants will increase in number, in their now familiar self-organizing fashion, for the foreseeable future and probably beyond.

The future will come, whatever mankind does. There are no experts on it. Therefore, a rational, and notably inexpensive way to build competence for whatever an acceptable future will demand is engagement in strategic foresight.[4]

References

Aeon (2016) The internet as an engine of liberation is an innocent fraud. Retrieved from https://aeon.co/essays/the-internet-as-an-engine-of-liberation-is-an-innocent-frau

Assange J (2016) WikiLeaks' Assange promises leaks on US election. Canadian Press, 4 Oct 2016. Retrieved from http://www.msn.com/en-ca/news/world/wikileaks-assange-promises-leaks-on-us-election-google/ar-BBwYBiG

BBC (2016) Swiss endorse new surveillance powers. BBC 25 Sept 2016. Retrieved from: http://www.bbc.com/news/world-europe-37465853

Bell S (2016) Canada's counter-radicalization efforts have 'little national coherence,' Public safety minister says. National Post, 14 Aug 2016. Retrieved from http://news.nationalpost.com/news/canadas-counter-radicalization-efforts-have-little-national-coherence-public-safety-minister-says

Buzan B (1987) An introduction to strategic studies: military technology and international relations. Palgrave Macmillan, London

[3]On 29 Jul 2016, John O. Brennan, the Director of the Central Intelligence Agency, made clear that while spying on each other's political institutions is fair game, making data public—in true or altered form—to influence an election is a new level of malicious activity, far different from ordinary spy versus spy maneuvers. See http://www.nytimes.com/2016/07/31/us/politics/us-wrestles-with-how-to-fight-back-against-cyberattacks.html?&moduleDetail=section-news-2&action=click&contentCollection=Politics&region=Footer&module=MoreInSection&version=WhatsNext&contentID=WhatsNext&pgtype=article.

[4]Foresight Canada (FC) definition of Strategic Foresight: The integrated capacity to see, think through and do what needs to be done NOW in the light of history-altering implications of the weak signals of change, while there is still time to act pro-actively and creatively and before hidden opportunities are lost and unseen threats have become crises.

Carr N (2016) Utopia is creepy: and other provocations. WW Norton & Company, New york
CBC News (2016) 'He knew I was pregnant.' Woman scrambling for housing after apartment scam. CBC News Toronto, 16 Sept 2016. Retrieved from: http://www.cbc.ca/news/canada/toronto/rental-scam-follow-1.3766908
Consumer Reports (2011) Your security. Consumer reports, June 2011
Cybersecurity Ventures (2016) Meet the hot cybersecurity companies to watch in 2016. Retrieved from http://cybersecurityventures.com/cybersecurity-500/
Economist (2014) Cyber-security: defending the digital frontier. The Economist, Special Report of 12th Jul 2014
Foreign Affairs (2016) ISIS' Virtual Puppeteers. Foreign Affairs Daily Post, 23 Sept 2016
Gibson W (1984) Neuromancer. Ace Book, New York
Harries D (2016) Scanning. Ottawa workshop for the Canadian Association of the Club of Rome: Climate Change Foresight, 8–9 Aug 2016
Kasulis K (2016) YouTube may be teaching someone to spy on you. Boston globe, 24 Sept 2016. Retrieved from: http://www.bostonglobe.com/ideas/2016/09/24/youtube-may-teaching-someone-spy
La Pointe C, Levin PL (2016) Automated war. Foreign Affairs Daily Post, 5 Sept 2016. Retrieved from: https://www.foreignaffairs.com/articles/2016-09-05/automated-war?cid=nlc-fatoday-20160907&sp_mid=52244007&sp_rid=amRzaGFycmllc0BiZWxsLm5ldAS2&spMailingID=52244007&spUserID=MjEwNDg3MDc3MTcwS0&spJobID=1001299689&spReportId=MTAwMTI5OTY4OQS2
MacFarquhar N (2016) A powerful russian weapon: the spread of false stories. NYT, 28 Aug 2016. Retrieved from: http://www.nytimes.com/2016/08/29/world/europe/russia-sweden-disinformation.html
McGill A (2016) Can Twitter fit inside the Library of congress? The Atlantic, 4 Aug 2016
Morgan S (2016) Cyber crime costs projected to reach $2 trillion by 2019. Retrieved from: http://www.forbes.com/sites/stevemorgan/2016/01/17/cyber-crime-costs-projected-to-reach-2-trillion-by-2019/#7ac44e733bb0
Nakashima E (2016) Powerful NSA hacking tools have been revealed online. Washington Post, 16 Aug 2016. Retrieved from: https://www.washingtonpost.com/world/national-security/powerful-nsa-hacking-tools-have-been-revealed-online/2016/08/16/bce4f974-63c7-11e6-96c0-37533479f3f5_story.html
New York Times (2016a) Snowden leaks Illegal but were 'a public service,' eric holder says, The New York Times, 1 Jun 2016. Retrieved from: http://www.nytimes.com/2016/06/01/us/holder-says-snowden-performed-a-public-service.html?action=click&contentCollection=U.S.&module=RelatedCoverage&region=EndOfArticle&pgtype=article
New York Times (2016b) Is democratic national committee email hacker a person or a russian front experts aren't? NYT 28 Jul 2016. Retrieved from: http://www.nytimes.com/2016/07/28/us/politics/is-dnc-email-hacker-a-person-or-a-russian-front-experts-arent
New York Times (2016c) US wrestles with how to fight back against cyberattacks. NYT, 31 Jul 2016. Retrieved from: http://www.nytimes.com/2016/07/31/us/politics/us-wrestles-with-how-to-fight-back-against-cyberattacks.html?&moduleDetail=section-news-2&action=click&contentCollection=Politics&region=Footer&module=MoreInSection&version=WhatsNext&contentID=WhatsNext&pgtype=article
New York Times (2016d) Hack of Democrats' Accounts was wider than believed, officials say, NYT 11 Aug 16. Retrieved from: http://www.nytimes.com/2016/08/11/us/politics/democratic-party-russia-hack-cyberattack.html?hp&action=click&pgtype=Homepage&clickSource=story-heading&module=first-column-region&region=top-news&WT.nav=top-news&_r=0
Perlroth N (2016) http://www.nytimes.com/2016/09/23/technology/yahoo-hackers.html?hp&action=click&pgtype=Homepage&clickSource=story-heading&module=first-column-region&region=top-news&WT.nav=top-news&_r=0
Ramzy A (2016) Australia stops online collection of census data after cyberattacks. NYT 10 Aug 2016. Retrieved from: http://www.nytimes.com/2016/08/11/world/australia/census-cyber-attack.html?ref=world

Ruiz RR (2016) Russian hackers leak U.S. star athletes' medical information. Retrieved from: http://www.theglobeandmail.com/sports/wada-claims-russian-cyber-espionage-group-has-hacked-its-systems/article31849291/

Sanger DE (2016) U.S. wrestles with how to fight back against cyberattacks, NYT 30 Jul 2016. Retrieved from: http://www.nytimes.com/2016/07/31/us/politics/us-wrestles-with-how-to-fight-back-against-cyberattacks.html?&moduleDetail=section-news-

Satter R, Cheslow D (2016) Apple issues iOS patch to thwart powerful spyware. Boston Globe, 25 Aug 2016. Retrieved from: http://www.bostonglobe.com/business/2016/08/25/apple-boosts-iphone-security-after-mideast-spyware-discovery/9vbGynNGtoCrFlJXVD0mOI/story.html?s_campaign=email_BG_TodaysHeadline&s_campaign=

Snowden E (2014) NYT 11 Aug 2016. Retrieved from: http://www.nytimes.com/2016/08/11/us/politics/democratic-party-russia-hack-cyberattack.html?hp&action=click&pgtype=Homepage&clickSource=story-heading&module=first-column-region&region=top-news&WT.nav=top-news&_r=0

Tsygankov A (2016) How cyber power fits into Russia's current foreign policy, russia direct, 3 Oct 2016. Retrieved from: http://www.russia-direct.org/profile/andrei-tsygankov

Tremonti AM (2016) The Current. 22 Aug 2016. Retrieved from: http://www.cbc.ca/radio/the current

Waterloo Chronicle (2016) Chinese in Canada feel chill of Beijing's reach. Retrieved from: http://m.waterloochronicle.ca/news-story/6827912-chinese-in-canada-feel-chill-of-beijing-s-reach

Weintraub P (2016) The world-wide cage. Retrieved from: https://aeon.co/essays/the-internet-as-an-engine-of-liberation-is-an-innocent-fraud

Author Biography

David Harries earned a Ph.D. in nuclear engineering from the University of London. He served in the Canadian military for several decades, was Director of Curriculum Planning and Deputy Commandant at the National Defence College of Canada, and has directed a MA program at the Royal Military College in Kingston. He has lived in 20 countries and paid working visits to another 93 ones. Six years (1996–2002) were spent based in Jakarta and Singapore and working throughout Asia. From 2004 to 2008 he ran one of the three MA programmes at the Royal Military College of Canada (RMC): *Security and Defence Management and Policy*. Educated as a nuclear engineer, he has worked in the public and private sectors as a senior military officer, as a consultant in personal and corporate security, and as a senior advisor and professor in heavy engineering, national development, humanitarian aid, post-conflict/post-disaster response and recovery, executive development and university education. He is currently based in Kingston, Canada from where he does research, curriculum development, teaching and facilitation of strategic foresight focusing on the family of five security domains, comparative civil-military relations, aboriginal entrepreneurship, leadership and leadingship, human security engineering and society resilience. Dr. Harries is most interested in the dynamics of the relationships among significant actors (individuals, organizations, and societies) and how they are governed by their biases, assumptions and interests as influenced by current events and trends. He is presently Chair of Canadian Pugwash Movement (Peace Nobel Price 1995) and head of its Foresight Committee, a Fellow of the World Academy of Art and Science, a member of the Board of Directors of the Global Initiatives Project and of ProteusCanada, head of the Leadership and Management community of IdeaConnector.net, Associate Executive Director of Foresight Canada and security foresight facilitator at Carleton University in Ottawa. David is interested in all five security domains, civil-military relations, aboriginal entrepreneurship, human security engineering, peacekeeping, and societal resilience.

A Conceptual and Legal Approach to the Cyberspace: The Dilemma Security Versus Freedom

Bernardino Cortijo

Abstract We should not assume that security and freedom are constantly at war with each other. When a person or a population is considered safe, do they consider themselves free? Usually, the opposite is true. If it is not so, what kind of freedom are we talking about? In fact, insecurity mainly has two key precursors; those with adverse natural, immediate or catastrophic effects, and those that arise from criminal or pre-criminal activities. The former naturally limit or condition us, as we try to limit or reduce their consequences, but we are not able to master them as such. But we cannot forget that there are other aspects which have complicated the picture. The use of computer tools in the Internet environment forces us to work from a much more technical point of view when we analyze crime materials and risk, as well as types of criminals and the use of tests and expert reports. Today, although the problems have not been solved, we have incorporated cyber intelligence activities, which allow us to access open sources via the Internet, or to enter private networks. These actions can be used both for good and for evil, but we have to consider and know them to have a slightly clearer idea of what we are really dealing with today when we are talking about cybercrime and cyber espionage. The current cyberspace world in which we live shows us the disadvantages we have in obtaining this freedom and, therefore, reaching equilibrium with safety, but they are not opposite concepts.

Keywords Cyberspace · Freedom · Security · Privacy

Submitted: 3.10.16; Accepted: 12.12.16.

B. Cortijo (✉)
Telefonica, Madrid, Spain
e-mail: dacor@dacorfdm.com

J.M. Ramírez and L.A. García-Segura (eds.), *Cyberspace*, Advanced Sciences and Technologies for Security Applications, DOI 10.1007/978-3-319-54975-0_3

1 Introduction: Equilibrium and Evolution of Crime and Technology

We should not assume that security and freedom are constantly at war with each other. When a person or a population is considered safe, do they consider themselves free? Usually, the opposite is true.

If it is not so, what kind of freedom are we talking about?

In fact, insecurity mainly has two key precursors; those with adverse natural, immediate or catastrophic effects, and those that arise from criminal or pre-criminal activities.

The former naturally limit or condition us, as we try to limit or reduce their consequences, but we are not able to master them as such.

Criminal activity has always had a process of development entwined with modern society at all times, with the means available, both for the citizens, companies and the State, and also for the criminals themselves, who are basically a reflection of society itself, since they are attempting to perform activities that are profitable or desirable or derived from current situations, but without following the legally permitted or assigned paths. It is frequently stated that criminals are always ahead of investigators, of justice, of laws; but they are not really ahead of society. Well, in the last 50 years, advances and technological developments have been so important in everyday life and in the global economy that this distance has been growing, for several reasons.

First, society's approach to the advanced technology that emerged with the development of processors and microprocessors that took place within the computer industry and the evolution of generic and specific software, initially within the business world, and later the domestic world, makes it very difficult for the courts to interpret each of the activities within the limit of the law. Going forward, legislators must devote more time in adapting to the types of technology, with more accurate knowledge of the true involvement in society of all technical and technological processes, primarily those related to the progress of personal computers, computer networks and data transmissions.

Here we have a multitude of documents and data of all kinds: confidential, official, records, reports, personal, games, utilities, taxes, personnel, budgets, etc. Criminal techniques are also being developed, such as crimes against intellectual property, software piracy, the copying of databases, and some fraudulent activities in the use of telecommunications, among others.

With the communications era already underway, connections through telephone cables were improving, with dedicated networks for large companies and banks, and communications through multiplexers and modems were helping to improve storage capacity. All this would culminate at the beginning of this century when the convergence of telecommunications for the transmission of information and data would be made via the INTERNET, merging computing technologies, more powerful personal computers, compression techniques, specific software, and communications themselves, such as a large network of computers linked together, with intermediate points specially designed with network servers and software that includes mathematical algorithms to link them and find the desired paths. All this required some rules and laws, which would naturally determine subsequent criminal investigation methods for us.

But that is not all. Software applied to the Internet, and its spread throughout the domestic and business world, in both medium and small businesses, opened new paths through global applications and is available to many, with the use of simulators that allow us to work remotely with applications and centralized resources, changing the means of communication at all levels. This includes the expanded use of email, communication channels such as chats, news feeds or digital panels, and the use of social networking applications.

But we cannot forget that there are other aspects which have complicated the picture. The use of computer tools in the Internet environment forces us to work from a much more technical point of view when we analyze crime materials and risk, as well as types of criminals and the use of tests and expert reports.

Today, although these problems have not been solved, we have incorporated cyber intelligence activities, which allow us to access open sources via the Internet, or to enter private networks. These actions can be used both for good and for evil, but we have to consider and know them to have a slightly clearer idea of what we are really dealing with today when we are talking about cybercrime and cyber espionage.

In this article, we will consider some of these aspects, as well as the difficulty for investigations, both in operations but especially for procedural and legal areas, concerning the modus operandi of these types of crime, defined according to the Spanish penal code and investigation processes.

The current cyberspace world in which we live shows us the disadvantages we have in obtaining this freedom and, therefore, reaching equilibrium with safety.

First came the stage of communications and applied data, and now we are in the stage of cyber-communications and cyber-data, altering the landscape of investigations in all criminal aspects, but especially with regard to terrorism and organized crime.

Matters such as money laundering or suspicious transactions, which until now could have been developed with local surveillance methods or the inspection of files and documents (both physical and digital), today have to delve into a cyber-group or a social network, in the deep web, a controlled prohibited forum with control and authentication systems.

But this is not enough; we have to exchange them with other open or closed sources, analyze by phases and in rings, encourage foresight, make reliable assumptions and execute defense options.

2 Examples of the Criminal Activities Involved. Terrorism and Organized Crime. Means Used to Commit Crimes

Let's begin with an example of a general crime, such as fraud, and more specifically, online and telephone fraud: we are not talking about the persecution of illegal contracts or the theft of terminals used later to commit a crime, but that there is a phase of discovery of new criminal activities, new technological methods, which vary day to day, controlling connections and international roaming, connecting organized crime networks together geographically, even connecting among themselves, through the execution of criminal acts (origin, roaming, final number) and, therefore, we must look toward methods and actions that will give us advance warning of the crimes.

International terrorism and organized activist groups, including organized criminal gangs, are using these means to commit major crimes, transferring the typical and traditional crimes to advanced techniques and technological means, in order to obtain anonymity and impunity, for themselves and for the criminal act itself.[1]

But these illicit activities, which particularly affect the well-being and the feeling and reality of security, are also those which disrupt social peace and cause the most serious incidents, leading society to allow intrusions into privacy and data online and in computers in order to prevent and prosecute these terrorist or organized crime activities relating to illicit traffic or other serious crimes.

Even in the cases where you can see that, in the defense of security, we are not limiting freedom, but acting to defend it, to allow citizens who respect the law to remain free, since the limitations of privacy or aspects of control will not entail any limitation for them.

Let's look at some examples. First, we have the use of secure encrypted point-to-point applications, inviolable software, octopus proxy servers (branches), confusion algorithms (using mathematical techniques to alter origins or IPs), IP havens, outdated legislation, international loopholes, "soft" uses of global alliances (EU, EUROPOL, etc.), which are tools easily and commonly used by these terrorists and organized crime groups to circumvent defense and investigation activities.

Second, we have attacks on institutions, the banking sector, communications, energy, embassies and especially consulates, governments, and attacks on shielded systems that control critical infrastructure, theft of personal or financial data, among others.

---

[1]Some examples about this technological anonymity evolution are the new Virtual Private Networks (VPN), such as Cyberghost VPN, Droid VPN, Free VPN 1.8, and The Onion Router (TOR). All of them are used to hide IP address and their respective origin. Also I2P (routers and tunnels), distributed and dynamic with encryption ElGamal, AES, ST, and PROXIES to hinder the investigations. Other options and tools are VPN clients routing through the Jon Do network (old JAP). For file sharing purposes, they use Freenet, TUVPN, WOT or Duck Duck Go finder.

In light of of this scenario, we must use investigation techniques based on technology and also on cyber intelligence.

3 Privacy, Data Use and Big Data

Let us now consider the problems that they offer. We shall discuss the FREEDOM-SAFETY dilemma, but in the scope of CYBER-FREEDOM—CYBER-SECURITY.

One of the main problems has already been detected by society in general. This is where the issue of privacy, the right to forget and the massive use of data (BIG DATA) appears. People need and want all current apps, and they need the massive use of their data and all the data that this entails, but they do not like to be controlled or potentially controlled.

If we are discussing the professional use and the fight against crime, investigation and security controls, there seems to be greater justification, because we are ensuring peace and order.

4 Right to Forget and Right to Remember

The right to forget has been heavily discussed (Alvarez Caro 2015; BOE 2015; Brock 2016; Chadwick 2014; Davara 2015; Touriño 2014), meaning that the Internet user is able to make decisions on the information and data that exists about themselves on the Internet and the links to such information.

Clearly whenever any data is included in any of the thousands of data servers that exist on the Internet, and every time information is related and managed from any of the search engines, it becomes even more difficult to ensure that the information will disappear or could disappear, even with the desire and the "right" to request such deletion. The reasons for the existence of such data may have also disappeared, or the circumstances may have changed, or it could be something personal, such as data or pictures, which does not make sense to keep after a certain period of time.

The reality is that while we all should consider this concept of privacy or possession of personal information, on the Internet it is not so simple. In fact, it should not be considered equally or have the same importance. It is something like a judge telling a jury not to consider a particular piece of evidence or testimony. But the result is very different.

And we can also discuss the right to remember, i.e. knowing things in order to make life easier for the upstanding citizen and to improve security, for which we also need access to alot of information. Today it is very important, because security has also become a right. Here there is another problem, the right to privacy, again opposed to the right to remember.

But let us look at a specific case.[2] Cyber-technology includes elements of communication, chips, GPS, potion, temperature, pressure, data and relationships. And applications include something much better, which is refined information reviewed by the individuals themselves and freely made available (Facebook, LinkedIn, etc.). When the person concerned and the owner "gives it all away", it would be wrong to think that all available means will not be used to protect it against an intruder or an attack from a criminal group.

5 Cyber-Criminal Controls and Impunity

Another issue is that there are controls for everything and everyone. There must be audit systems to help recognize, albeit a posteriori, people who commit misdemeanors.

We are living in a cyber-world and organized criminals and terrorists have: impunity, anonymity (Salvador Carrasco 2012), speed, geographic freedom, simple relationships between themselves, etc. That's why we question the appearance of controls to facilitate global or shared security, thereby avoiding impunity as far as possible. This will appear again in such popular topics as the use of public Internet access, essentially free or open WiFi hotspots that do not require any identification, such as those provided in public places, city halls, bars, coffee shops, etc.

6 Cyber Intelligence

To counteract all this criminal activity, we need to use new and more sophisticated techniques which are forward-looking and highly advanced. The use of cyber intelligence and technological measures (facial recognition, digital control, voice control, recognition of proxy use and ports, source access, social networking destination, chats and emails, etc.).

Therefore, we must find the reasonable limit of freedom and especially of cyber-freedom in order to offer security, just as the limit of privacy should be maintained to offer required services and improve everyone's life.

In any case, we must differentiate these indicators and sensors that allow for predefined perspectives on new techniques based on computer analysis of cyber intelligence, following the cycle of intelligence and mapping out indicators that suit our needs.

---

[2]Some examples include: Endomondo, iDoctus, Medisafe, SIG: OruxMaps, QGis, Collector for ArcGIS, Google Maps, OSM, Fulcrum, Waze Social, Couple, Cozi, Facebook, Pplkpr, Swiftkey, Amazon; as well as Magnetoresistive random-access memory.

We are looking at the right to privacy against the right to knowledge, and not information, where the right to avoid fraud prevails over the right to forget a paid debt. Yet I am not entirely sure whether they are applicable or admissible, although in the case of the fight against terrorism and organized crime, which can destabilize territories, it seems clear that the means are necessary and the damage is very small compared to the possible consequences.

Technology provides more data and more solutions, not only through cyber intelligence, but also through cyber data and Big data. We must emphasize that location is no longer relevant, as the cloud allows users to be connected at all times. Our cyber-freedom should no longer involve safeguarding information, but ensuring that it is not used inappropriately. And when there is a "greater evil", such as terrorism and money laundering, there is no choice but to use these online techniques.

7 Conclusion

We should reflect on freedom, security and rights. More specifically, we think that:

1. Security is insecurity where security does not exist. And if this happens, there is no freedom.
2. To meet our online needs we want and need globalized and personalized information, but we do not want our data drifting around in cyberspace.
3. The right to forget is recognized by many, but we all use the right to remember.
4. In the virtual world of communications, we must deeply respect the rights linked to freedom, but also to cyber-freedom: the power to decide on existing personal data.
5. Another consideration is the cyber freedom of individuals versus collective or social cyber freedom, which should be demanded to provide information and data on social networks and online means of communication.

Who has more rights? Should society know and collect relevant data on its citizens, or should citizens safeguard this data, even though it could be important for the common good, the social good or security in general to have it?

Cyberspace is a unique medium, and should be treated as such. But social freedom, collective freedom and social peace should be given priority over individual rights. Nevertheless, we must put some limits on the use of media and technologies which are not strictly necessary for the common good.

We now have increasingly advanced technical and technological resources to perform inspections, store information, segregate said information, find and link data, and finally know even more than the owners or interested parties themselves.

In any case, this should not be an obstacle to progress and development, but it should encourage, in the broadest sense, legislators to be able to verify and encourage proper use, while also dealing with the bad apples who abuse these technological and temporal possibilities.

References

Alvarez Caro M (2015) El Derecho al Olvido en Internet. Editorial REUS
BBC (2006) Magnetic memory chip unveiled. BBC News, 07 Oct 2006
BOE (2015) Código del Derecho al Olvido. BOLETIN OFICIAL DEL ESTADO (last versión 2017/24/01)
Brock G (2016) Right to be forgotten. George Brock I.B., Tauris
Bueno F (2012) FODERTICS estudios sobre Derecho y nuevas Tecnologías. Editorial Andavira
Chadwick R (2014) The right to know and the right not to know. Cambridge University Press, Cambridge
Davara MA (2015) Manual de Derecho Informático Thomson Reuters Aranzadi, 11 edn
De Salvador Carrasco L (2012) Redes de Anonimización en Internet: Cómo funcionan y cuándo son sus límites. Revista Instituto Español de Estudios Estratégicos N.16 of 2012 ieee.es
Touriño, A. (2014). El Derecho al Olvido y a la Intimidad en Internet. La Catarata

Author Biography

Bernardino Cortijo Former National Police Commissioner, Mr. Cortijo has extensive experience in Information and Cyber-related security. He was Vice-President of Security in Terra Lycos, as well as Head of the Central Police Cyber crimes Unit in Madrid, Spain. Mr. Cortijo also holds several graduate and postgraduate degrees including a Master's Degree in Internet and Telecommunications.

The Digital Revolution in Developing Countries: Brief Analysis of the Dominican Republic

Luis A. García-Segura and Juan Cayón Peña

Abstract Information and Communications Technologies for Development (ICT4D) can help people in developing countries pursue a wider variety of economic and social activities, which in the majority of the cases can reduce overall poverty significantly. As of this moment, there is an important "Digital divide" regarding broadband internet access between Developed and Developing countries, calling for immediate action plans around the world to build the digital infrastructure that can foster digital economies and thus the digital revolution. Applied to sustainable development, the data revolution calls for the integration of this new data with traditional data to produce high-quality information that is more detailed, timely and relevant for many purposes and users, especially to foster and monitor sustainable development. On the infrastructure side, broadband access must continue to be a priority in every digital revolution strategy, given the new Digital divide gaps that have emerged in the past ten years. On the social side, Big data for development can be one of the premier mechanisms that drives the data revolution within the bigger digital revolution picture.

Keywords Cyberspace · ICT4D · Big data · Digital revolution · Data revolution · Dominican Republic

Submitted: 1.9.16; Accepted: 29.12.16.

L.A. García-Segura (✉) · J.C. Peña
Nebrija-Santander Chair on Risk and Conflict Management, Nebrija University, Madrid, Spain
e-mail: lgarcise@nebrija.es

J.C. Peña
e-mail: jcayon@nebrija.es

J.M. Ramírez and L.A. García-Segura (eds.), *Cyberspace*, Advanced Sciences and Technologies for Security Applications, DOI 10.1007/978-3-319-54975-0_4

1 Introduction

Cyberspace is a fascinating medium that is transforming our way of life at a staggering rate and pace. Developed countries have been quick to reap the benefits of the Digital ecosystem which the Internet has catapulted. In contrast, the majority of developing countries have not followed in the steps of their developed counterparts, and have yet to experience the full transformative capacity that the Digital ecosystem has to offer. This paper briefly analyzes the way in which developing countries can transform their economies in order to accelerate their social and economic transitions into fully Digital economies.

Let us start by reflecting on the present role that Information and Communications Technologies (ICTs) have to play in the Development arena all together.

2 ICTs and Development

2.1 Human Development

In 1986, the United Nations General Assembly (UNGA), in its resolution 41/128, adopted the Declaration on the Right to Development. Its article 1 states the following:

> The right to development is an inalienable human right by virtue of which every human person and all peoples are entitled to participate in, contribute to, and enjoy economic, social, cultural and political development, in which all human rights and fundamental freedoms can be fully realized (UNGA 1986, Article 1.1).

Ever since this Declaration was adopted, the Right to Development has been one of the main topics of research and work of the United Nations Development Programme (UNDP).[1] Human development as understood by the UNDP (2015) calls for the expansion of the richness of human life, rather than simply the richness of the economy in which human beings live. The dimensions this expansion calls for are highlighted in the following diagram (Fig. 1).

In order to measure the advancements of this dimensions, the UNDP created the Human Development Index (HDI) as a:

> summary measure of average achievement in key dimensions of human development: a long and healthy life, being knowledgeable and have a decent standard of living. The HDI is the geometric mean of normalized indices for each of the three dimensions (UNDP 2016, p. 1).

[1]The UNDP currently works in nearly 170 countries and territories, helping to achieve the eradication of poverty, and the reduction of inequalities and exclusion. They help countries to develop policies, leadership skills, partnering abilities, institutional capabilities and build resilience in order to sustain development results. For more information visit: http://www.undp.org/content/undp/en/home/operations/about_us.html.

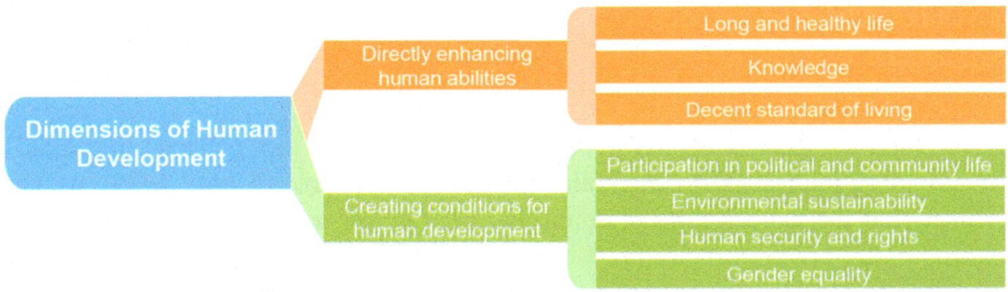

Fig. 1 UNDP (2015, p. 1)

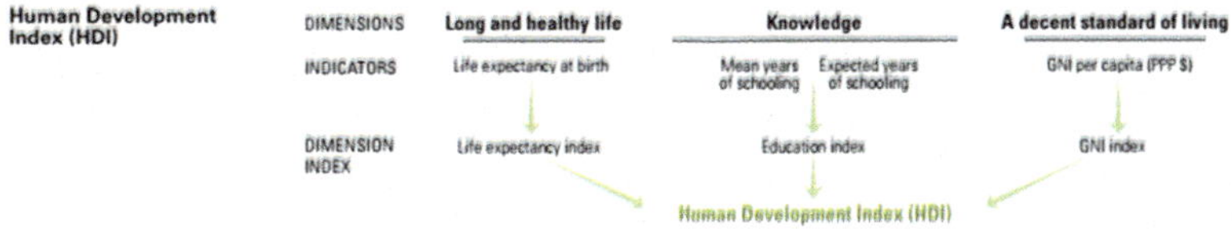

Fig. 2 UNDP (2016, p. 1)

It can be clarified in the following diagram (Fig. 2).

We strongly believe, as well as other academics do,[2] in the impact that ICTs can have in advancing/improving every one of these dimensions. In the past 20 years, a significant amount of academic and scientific evidence has been published,[3] that links Human Development progression with the strategic implementation of ICTs in all aspects of human life.

For example, if we take the top ten countries from the HDI 2015, the majority of them are in Europe and North America (UNDP 2016):

1. Norway, with a top score of 0.944.
2. Australia
3. Switzerland
4. Denmark
5. Netherlands
6. Germany
7. Ireland

[2]The works of Professor Richard Heeks (cited in this text) and the Centre for Development Informatics of the University of Manchester is focused on the role of ICTs in socio-economic development. Some of their current research topics include: Digital economy; Digital transformation; Digital inclusion; Digital sustainability and Digital theory. For more information visit: http://www.cdi.manchester.ac.uk/research/.

[3]See the publications sections of: Economic Commission for Latin America and the Caribbean (CEPAL) (http://www.cepal.org/en/publications); World Economic Forum (https://www.weforum.org/reports/); OECD (http://www.oecd.org/publications/); Inter-American Development Bank (https://data.iadb.org/) and the International Telecommunications Union (http://www.itu.int/en/publications/Pages/default.aspx).

8. United States of America
9. Canada
10. New Zealand.

If we then compare this group of countries with the top nations in the ICT Development Index[4] (IDI) (International Telecommunication Union 2016), we can also see that the majority of the countries are from the European continent:

1. Korea (Rep.)
2. Iceland
3. Denmark
4. Switzerland
5. United Kingdom
6. Hong Kong, China
7. Sweden
8. Netherlands
9. Norway
10. Japan.

Furthermore, in the Networked Readiness Index (NRI)[5] published by the World Economic Forum (Baller et al. 2016), Europe remains at the technology frontier with 7 out of the top 10 NRI countries coming from this continent (Finland, Sweden, Norway, Netherlands, Switzerland, United Kingdom and Luxembourg).

On the other hand, the bottom ten countries of the HDI, IDI and NRI where:

HDI (United Nations Development Programme 2016)	IDI (International Telecommunication Union 2016)	NRI (Baller et al. 2016)
1. Niger, with a worst score of 0.348 2. Central African Republic 3. Eritrea 4. Chad 5. Burundi	1. Niger 2. Chad 3. Guinea 4. South Sudan 5. Burundi 6. Congo (Dem. Rep.)	1. Chad 2. Burundi 3. Haiti 4. Mauritania 5. Madagascar 6. Guinea

(continued)

[4]The ICT Development Index (IDI), which has been published annually since 2009, is a composite index that combines 11 indicators into one benchmark measure. It is used to monitor and compare developments in information and communication technology (ICT) between countries and over time. Some of the indicators of the IDI include: International Internet bandwidth (bits/s) per Internet user; Percentage of individuals using the Internet and Adult literacy rate. For more information visit: http://www.itu.int/en/ITU-D/Statistics/Pages/publications/mis2015/methodology.aspx.

[5]The NRI is a composite indicator made up of four main categories (sub-indexes), 10 subcategories (pillars), and 53 individual indicators distributed across these main categories: Environment (Political, regulatory, business and innovation environment); Readiness (Infrastructure, affordability and skills); Usage (Individual, business and government) and Impact (Economic and social) (Baller et al. 2016, p. xi).

(continued)

HDI (United Nations Development Programme 2016)	IDI (International Telecommunication Union 2016)	NRI (Baller et al. 2016)
6. Burkina Faso 7. Guinea 8. Sierra Leone 9. Mozambique 10. Mali	7. Ethiopia 8. Malawi 9. Tanzania 10. Madagascar	7. Myanmar 8. Malawi 9. Nicaragua 10. Liberia

As we can see, almost all of the countries in the worst positions are from Africa, followed by Asia and Latin America. Therefore, it is safe to say that there is a direct link between Human Development and ICT implementation in society, and that it can even help reduce poverty levels.

Accordingly, Heeks (2014b, p. 9) points out three different perspectives on poverty eradication that can be applied within an ICT for Development (ICT4D) strategy:

1. Economic: seeing income generation as the route to poverty eradication.
2. Livelihoods: seeing poverty eradication as deriving from delivery of a variety of livelihood assets; not just money but also health, skills, information, etc.
3. Capabilities: seeing poverty eradication deriving from development of greater freedom to act, encapsulated by the notion of "roles".

Out of these three approaches, we think that the capabilities approach is best suited to measure the impact of poverty eradication within ICT4D strategies, projects and plans in the years to come. This capabilities approach draws on the works of Sen (1985), and defines poverty as a lack of abilities and chances to do what is necessary to progress in life; whereas development is the expansion of individual freedoms so that a person can pursuit whatever goals or values they regard as important, especially economic and social activities.

As a result, we think that ICT4D can help people in developing countries pursue a wider variety of economic and social activities, which in the majority of the cases can reduce overall poverty significantly.

2.2 *From the MDGs to the SDGs*

In the past two decades, we have witnessed the debate, research and implementation of programs, plans, objectives and strategies aimed at, among other overarching goals, to drastically reduce extreme hunger and poverty. Perhaps the most important initiative up to the year 2015 was the Millennium Development Goals (MDGs) plan, which where a series of time-bound targets with a deadline of 2015, arising from the UN Millennium Declaration adopted in September 2000. These targets looked to address extreme poverty in a series of dimensions, as can be seen in the names of the goals themselves[6]:

1. Eradicate extreme hunger and poverty
2. Achieve universal primary education
3. Promote gender quality and empower women
4. Reduce child mortality
5. Improve maternal health
6. Combat HIV/AIDS, Malaria and other diseases
7. Ensure environmental sustainability
8. Develop a global partnership for Development.

These goals were the reflection of the critical problems affecting undeveloped countries in comparison with developed countries, and became for the past 15 years the overarching development framework for the world. According to the final report published last year by the United Nations (2015),

> the world community has reason to celebrate…{because} the MDGs have saved the lives of millions and improved conditions for many more (p. 4). Some of the achievements that have improved the conditions for millions of people include (UN 2015, pp. 4–7):

1. Globally, the number of people living in extreme poverty has declined by more than half, falling from 1.9 billion in 1990 to 836 million in 2015. Most progress has occurred since 2000.
2. The primary school net enrolment rate in the developing regions has reached 91% in 2015, up from 83% in 2000.
3. Women now make up 41% of paid workers outside the agricultural sector, an increase from 35% in 1990.
4. Despite population growth in the developing regions, the number of deaths of children under five has declined from 12.7 million in 1990 to almost 6 million in 2015 globally.
5. Since 1990, the maternal mortality ratio has declined by 45% worldwide, and most of the reduction has occurred since 2000.
6. Over 6.2 million malaria deaths have been averted between 2000 and 2015, primarily of children under five years of age in sub-Saharan Africa. The global malaria incidence rate has fallen by an estimated 37% and the mortality rate by 58%.
7. Ozone-depleting substances have been virtually eliminated since 1990, and the ozone layer is expected to recover by the middle of this century.
8. Official development assistance from developed countries increased by 66% in real terms between 2000 and 2014, reaching $135.2 billion.

---

[6]Each goal was assigned a series of targets and indicators which can be consulted at: http://www.unmillenniumproject.org/goals/gti.htm.

In relation to ICT4D, Goal number 8, to "Develop a global partnership for Development" included the following target and indicators regarding ICTs Millennium Project (2006):

> Target 18. In cooperation with the private sector, make available the benefits of new technologies, especially information and communications technologies.
> Indicators:
> 47. Telephone lines and cellular subscribers per 100 population (ITU).
> 48. Personal computers in use per 100 population and Internet users per 100 population (ITU).

As we can see, the target was vague and the indicators only took into consideration technical milestones. Therefore, there was no real systematic approach to help foster development from an abilities standpoint.

Even though a lot of progress was made, there is still a significant part of the indicators and targets set forth by the MDGs which were not completed. Particularly, the UNGA (2015, p. 5) voiced the following challenges facing our world today:

1. Billions of people continue to live in poverty;
2. Enormous disparities of opportunity, wealth and power within and among countries;
3. Youth unemployment is a major concern;
4. More frequent and intense natural disasters;
5. Natural resource depletion and adverse impacts of environmental degradation;
6. Violent extremism, terrorism and related humanitarian crises.

Taking into account all of the above problems, the Economic Commission for Latin America and the Caribbean (CEPAL) (2013b, p. 97) concluded in 2013 that in order to make progress on the structural change that is necessary for development and the reduction of inequality in the region, the strategy for the future requires formulating and implementing a new phase of ICT strategies based on the integrated development of the digital economy and the network of economic and social activities facilitated by the Internet.

Therefore, in order to face these challenges and at the same time try to complete the MDGs, a new development Agenda was adopted by UN leaders for the next 15 years: the "2030 Agenda for Sustainable Development", adopted by the General Assembly on September 25th 2015 (UNGA 2015). This new plan created a set of 17 "Sustainable Development Goals" (SDGs) and 169 targets to stimulate action in areas of critical importance for humanity and the planet. The 17 SDGs are the following:

Goal 1. End poverty in all its forms everywhere.
Goal 2. End hunger, achieve food security and improved nutrition and promote sustainable agriculture.
Goal 3. Ensure healthy lives and promote well-being for all at all ages.
Goal 4. Ensure inclusive and equitable quality education and promote lifelong learning opportunities for all.

Goal 5. Achieve gender equality and empower all women and girls
Goal 6. Ensure availability and sustainable management of water and sanitation for all.
Goal 7. Ensure access to affordable, reliable, sustainable and modern energy for all.
Goal 8. Promote sustained, inclusive and sustainable economic growth, full and productive employment and decent work for all.
Goal 9. Build resilient infrastructure, promote inclusive and sustainable industrialization and foster innovation.
Goal 10. Reduce inequality within and among countries.
Goal 11. Make cities and human settlements inclusive, safe, resilient and sustainable.
Goal 12. Ensure sustainable consumption and production patterns.
Goal 13. Take urgent action to combat climate change and its impacts.
Goal 14. Conserve and sustainably use the oceans, seas and marine resources for sustainable development.
Goal 15. Protect, restore and promote sustainable use of terrestrial ecosystems, sustainably manage forests, combat desertification, and halt and reverse land degradation and halt biodiversity loss.
Goal 16. Promote peaceful and inclusive societies for sustainable development, provide access to justice for all and build effective, accountable and inclusive institutions at all levels.
Goal 17. Strengthen the means of implementation and revitalize the Global Partnership for Sustainable Development.

Before mentioning the ICT component of the MDGs, we would like to point out highlight the most frequently quoted definition of Sustainable development[7]:

> Sustainable development is development that meets the needs of the present without compromising the ability of future generations to meet their own needs. It contains within it two key concepts: the concept of needs, in particular the essential needs of the world's poor, to which overriding priority should be given; and the idea of limitations imposed by the state of technology and social organization on the environment's ability to meet present and future needs (International Institute for Sustainable Development 2016, p. 1).

Although this definition may seem clear to us, we must advise that the term is an evolving and disputed concept, making it hard sometimes to find a universal understanding of it.[8] Nonetheless, it was used to name the development goals of the

[7]Definition taken from the report "Our Common Future", drafted by the World Commission on Environment and Development, also known as the Brundtland Commission.

[8]Regarding the definition of this term, Berger-Walliser and Shrivastava (2015) stated that: "There is no authoritative legal definition of sustainable development. The absence of a clear definition is sometimes seen as a contributing factor to the delay in effectively addressing environmental, social, and economic concerns. The word sustainability or sustainable development is ubiquitous, and like any overused term, is in danger of being watered down, misused, abused, and losing its original meaning. Various approaches to sustainable development exist and different communities employ the term with different meanings depending on the conditions and settings of where it is used (p. 422)".

United Nations and we think it summarizes part of the central concerns of many citizens all over the world.

For instance, the OECD (2016b) has recognized the SDGs as being the most ambitious, diverse and universal development roadmap in history, requiring investment needs in developing countries of an estimated USD 3.3–4.5 trillion per year, a figure well beyond the approximately USD 132 billion provided as official development assistance in 2015. Under these circumstances, the amount of money, which will be allocated for ICT4D projects globally will be remarkable, most of it coming from Developed countries and UN agencies.

Consequently, the United Nations Economic and Social Council (UNESC) (2014, p. 3) has stated that there are three ICT capabilities that are especially important for economic and social development:

1. Enabling greater efficiency in economic and social processes;
2. Enhancing the effectiveness of cooperation between different stakeholders;
3. Increasing the volume and range of information available to people, businesses and Governments.

These capabilities are present in the following targets and projects proposed by the SDGs (UNGA 2015):

> Build resilient infrastructure, promote inclusive and sustainable industrialization and foster innovation, target 9.c: Significantly increase access to information and communications technology and strive to provide universal and affordable access to the Internet in least developed countries by 2020 (Goal 9, p. 21).
>
> Strengthen the means of implementation and revitalize the Global Partnership for Sustainable Development, target 17.8: Fully operationalize the technology bank and science, technology and innovation capacity-building mechanism for least developed countries by 2017 and enhance the use of enabling technology, in particular information and communications technology (Goal 17, p. 26).

The launch of a Technology Facilitation Mechanism (TFM): in order to support the SDGs, a multi-stakeholder collaboration hub between Member states, civil society, the private sector, the scientific community, United Nations entities and other stakeholders. The hub includes these mechanisms:

1. A United Nations inter-agency task team on science, technology and innovation for SDGs.
2. An online platform to facilitate access to information, knowledge and experience on science, technology and innovation facilitation initiatives and policies.
3. A multi-stakeholder forum on science, technology and innovation for SDGs (p. 30).

Ultimately, all of these goals, targets and projects call for a digital revolution that can advance each one of the SDGs. That being the case, we will now explain the main objectives and points of this Revolution.

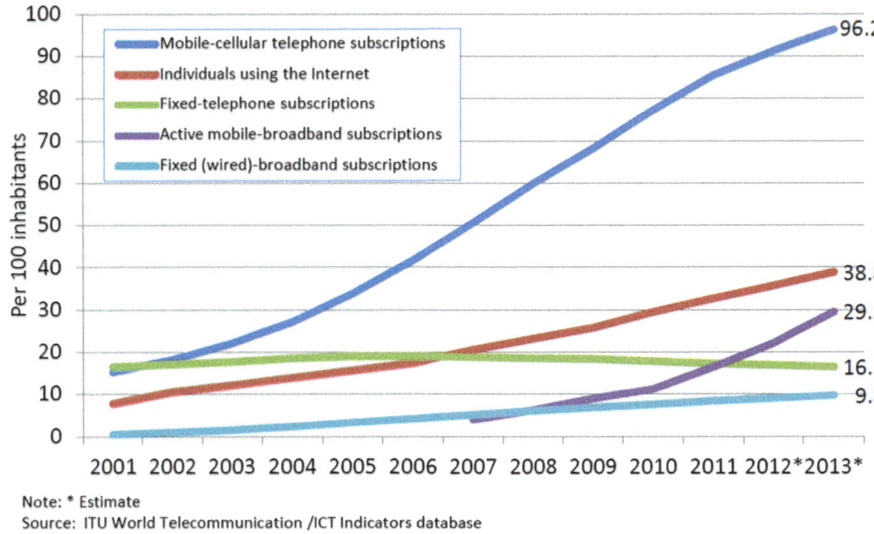

Fig. 3 Global ICT developements. Heeks (2014a, p. 21)

3 The so Called "Digital Revolution"

3.1 Digital Divide

In the following graphic we can see the growth evolution of the principle ICT instruments in the world's population (Fig. 3).

Clearly, mobile-cellular subscriptions have had a spectacular penetration rate, due in part to cheap phones and diverse service plans that include internet access.[9] Nevertheless, as we will focus on later in this paper, most of the objectives for the digital revolution involve fixed broadband internet subscriptions, which are still

[9]In a 2014 article titled "The rise of the cheap smartphone", The Economist (2014) stated "In both rich countries and poor ones, cheaper smartphone brands are making inroads. Demand for pricey phones, mainly in developed economies, is slowing, but that for less expensive devices is booming. People buying their first smartphones today, perhaps to replace a basic handset, care less about the brand and more about price than the richer, keener types of a few years ago". A year later, the same publication concluded that smartphones "…have become the fastest-selling gadgets in history, outstripping the growth of the simple mobile phones that preceded them. They outsell personal computers four to one. Today about half the adult population owns a smartphone; by 2020, 80% will. Smartphones have also penetrated every aspect of daily life. The average American is buried in one for over two hours every day. Asked which media they would miss most, British teenagers pick mobile devices over TV sets, PCs and games consoles. Nearly 80% of smartphone-owners check messages, news or other services within 15 min of getting up. About 10% admit to having used the gadget during sex (The Economist 2015, p. 1)".

very low in Developing countries. Regarding broadband access, a recent report by the OECD and the Interamerican Development Bank (IDB) (2016) concluded the following:

> Broadband networks are the foundation of digital economies. Increased availability and effective use of the services enabled by broadband can advance social inclusion, productivity and good governance. A range of challenges has to be overcome, however, in providing readily accessible, universal and locally relevant broadband-based services in many parts of the world (OECD & IDB 2016, p. 15).

As of this moment, there is an important "Digital divide" regarding broadband internet access between developed and developing countries,[10] calling for immediate action plans around the world to build the digital infrastructure that can foster digital economies and thus the digital Revolution. Concerning this problem, a key policy focus of the ICT4D approach has been the Digital divide, with considerable emphasis on those who are not digitally connected. However, it has become increasingly evident that simply focusing on access misses the point. As more areas and people become digitally connected, new divides are emerging in terms of capabilities and resources, due to the rapid change in technology and the highly innovative nature of new ICT applications (UNESC 2015, p. 12).

As a result, ICTs can create and also exacerbate the inequalities that exist in society. Further, inadequate availability of appropriate local content on the Internet can hinder inclusive[11] digital development (UNESC 2015, p. 18). In a like manner, Heeks (2014b) argues that once the barriers of ICT access have been overcome, recipients must have the skills, knowledge, money and motivation to create real and lasting development results. This is explained in the following graphic (Fig. 4).

The skills, knowledge and learning aspects of the Information Impact Chain presented by Heeks (2014b) can be summarized in one word: Education.

---

[10]The United Nations General Assembly Economic and Social Council (2015, p. 22) stated this regarding the emergence of new Digital divides: "While divides in access to basic communications between and in countries have been diminishing, new divides have been growing in access to broadband networks and the services that they enable. Particular concern has been expressed that least developed countries may fall behind developed and other developing countries in broadband access and use, that rural areas are often disadvantaged in comparison with urban areas and that there remains a gender gap in ICT access and use".

[11]The UNESC (2014, p. 6) stated the following regarding inclusiveness and the Digital divide problem: "The challenge of inclusiveness has been at the heart of ICT4D policymaking. Developed countries have better ICT infrastructure, enjoy more pervasive ICT usage, and gain earlier access to ICT innovations than developing countries. Urban areas and wealthier social groups in developing countries are similarly advantaged over rural areas and poorer communities. ICT access and use are less prevalent in groups that are socially or economically marginalized, such as women, youth, unqualified or subsistence workers, ethnic minorities and those with special needs or disabilities. While the value of ICTs to all is recognized, therefore, its benefits may accrue disproportionately within society".

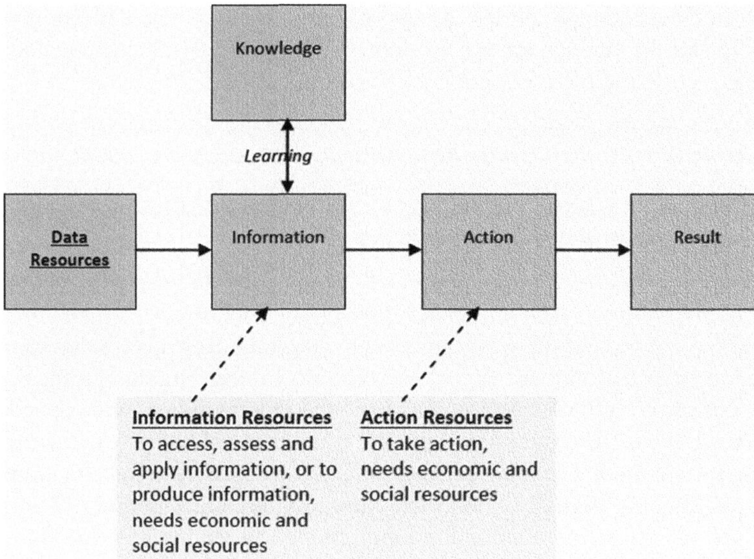

Fig. 4 The information impact Chaon. Heeks (2014b, p. 18)

3.2 *Education*

Access to ICT alone does not guarantee reaping its full benefits. It is necessary to climb up the digital "role ladder", that is to say, to make the transition from being active users of e-mail and Internet applications to becoming producers, entrepreneurs and innovators, taking advantage of ICT-enabled opportunities. Climbing up this ladder would require the development of capabilities—technological and other skills such as critical and analytical thinking, problem solving and creativity. Although there is a surge in the number of millennials from developing countries who are active Internet users, very few end up climbing up the role ladder to become producers, entrepreneurs or innovators, as they lack the necessary digital capabilities and enabling opportunities (UNESC 2015, pp. 7–8).

For example, in the past decade, most of the countries in the region have made significant progress in the incorporation of digital technology in education systems, especially in terms of access and infrastructure. However, progress is still weak on two key factors associated with effective ICT use: the uptake of the technology by the user and the development of relevant educational content (CEPAL 2013b, p. 100).

Correspondingly, the OECD (2016a) recently identified how recent technological trends are affecting education worldwide:

(1) Information and data

1. Teaching students and teachers how to evaluate the validity of online information
2. Developing strategies to combat plagiarism
3. Providing training on the use of Big data.

(2) Learning and teaching

1. Integrating technology into the classroom
2. Utilizing collaborative learning platforms to share and expand knowledge
3. Teaching programming and advanced computing skills
4. Implementing self-paced and personalized e-learning.

(3) Cyber-risks

1. Teaching students and parents how to protect themselves from online risks
2. Strategies to combat cyberbullying
3. Developing protocols to protect sensitive data from security flaws and hacking.

The actions needed to support the first two trend groups mainly have to come from the government side, being that they affect in most part the public education system. For example, the integration of technology into the classroom must begin in the early stages of the education system in order to provide the minimum ICT literacy skills that will be needed throughout the academic life of any student. Once this is done, then students and teachers can start to use collaborative platforms to share and expand knowledge as well as to implement self-paced and personalized e-learning, which by themselves are crucial skills that will aid the lifelong learning process[12] that students must embark on today.

For this reason, the human infrastructure needs to be strengthened in terms of ICT production and consumption capabilities, including complementary capabilities such as communication, and interpersonal and analytical skills. These complementary capabilities are essential in converting ICT skills to productive employment opportunities.[13] This calls for educational policy interventions at

[12]One of the UNESCO's strategic objectives for education for the 2014–2021 period is "Supporting Member States to develop education systems to foster high quality and inclusive lifelong learning for all (UNESCO 2014, p. 31)". Furthermore, UNESCO states that "The lifelong learning paradigm has been framed and promoted by UNESCO since the 1970s. If education is to respond effectively to challenges posed by the rapid and constant changes which characterize the 21st century and its development, learning will have to take place throughout life (UNESCO 2014, p. 33)".

[13]About the impact of technology in labour and employment, the UNESC (2016, p. 3) stated this year that: "Technological change is not neutral; it can favour either labour or capital. Technological change is, in essence, disruptive and, in the short term, creates winners and losers.

primary, secondary and tertiary educational and vocational training levels. Focus should also be on how to empower people to emerge as innovators and producers of ICT applications suited for specific local needs (UNESC 2015, p. 13).

In order to empower people to emerge as innovators and producers of ICT applications, we advocate the development of a true data revolution, as a core part of the digital revolution which we have mentioned before.

3.3 *Data Revolution*

The High-Level Panel of Eminent Persons on the Post-2015 Development Agenda (2013) stated the following about the data revolution:

> We also call for a data revolution for sustainable development, with a new international initiative to improve the quality of statistics and information available to people and governments. We should actively take advantage of new technology, crowd sourcing, and improved connectivity to empower people with information on the progress towards the targets (Development Agenda 2013, p. 22).

Since ICTs have increased their presence in developing countries in terms of reach, scope and depth, we are faced with the emergence of a digital "nervous system" for development. It is pervasive digital infrastructure in which most development organizations, from international agencies to government departments and small community-based organizations, have Internet access and in which a large proportion of individuals in developing countries have digital mobile phone access (UNESC 2015).

Henceforth, applied to sustainable development, the data revolution calls for the integration of this new data with traditional data to produce high-quality information that is more detailed, timely and relevant for many purposes and users, especially to foster and monitor sustainable development (United Nations Secretary-General 2014).

In the following graphic of the Digital economy environment, the data revolution is characterized as a technological platform via the Big data process (Fig. 5).

This characterization of the data revolution through Big data has been sustained by the experts in the past years. Primarily, the United Nations Economic and Social Council (2014) concluded that there are five emerging trends that have the potential to drive further changes in the relationship between ICTs and development beyond 2015:

(Footnote 13 continued)

While disruptive technologies will be critical to a transformation towards sustainable development, their benefits may disproportionally go to people in the countries that innovate or to a small fraction of the population".

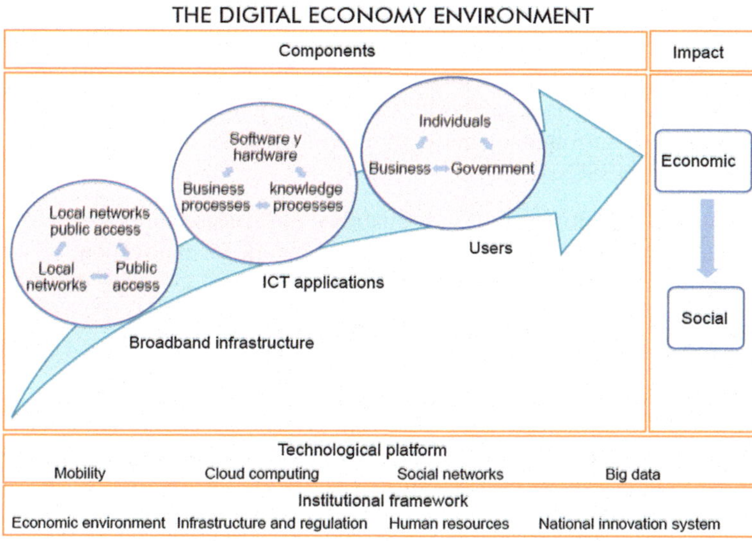

Fig. 5 CEPAL (2013b, p. 10)

1. Datafication of business and government organization and practice;
2. The emergence of Big data and Big data analysis as new resources for understanding social and economic processes;
3. Widespread adoption of cloud computing;
4. The emergence of the Internet of things;
5. The deployment of smart systems to improve efficiency and productivity throughout economies.

Moreover, Heeks (2014a) stated that there are three dominant aspects to a development data revolution:

1. Big development data: the emergence of very large datasets relating to phenomena within developing countries.
2. Open development data: the greater availability of developing country datasets for general use.
3. Real-time development data: the availability of developing country data in real time; that is, simultaneous to the moment of the data-creating event.

Because of this outlook, business and particularly government organization and practice can greatly improve their respective performance on a wide scale. Take for example, the access to public services in remote rural areas. Many people in developing countries can be excluded of some public services due to the fact that they do not have the means and/or time to travel to the district headquarters where some of these services might be administered. For these people, ICTs can help overcome some of these issues by reaching out to low-income communities though e-government and m-government applications. According to the UNESC (2015), evidence suggests these applications can improve consistency and citizen satisfaction; as well as significantly reduce corruption.[14]

Nonetheless, in order to implement effectively any of the data revolution mechanisms aforementioned, authorities must take into account the principal contextual constraints surrounding the deployment of ICTs in developing countries, which include (UNESC 2014):

1. The availability, affordability and reliability of infrastructure;
2. The quality of the legal and regulatory framework for innovation;
3. The human and institutional capabilities needed to leverage developmental value from programmes and projects;
4. The financial resources for investment in infrastructure, human capacity and operational costs.

In addition to these four constraints, we think that ICTs have to be adapted to local contexts and culture in order to have the necessary transformative impact needed to achieve the goals set forth by the MDGs (UNESC 2016) (Centre for International Governance Innovation and The Royal Institute for International Affairs 2016, p. viii). For this reason, with the help of some current examples, we will now focus our attention on a specific developing country: The Dominican Republic.

4 The Case of the Dominican Republic

4.1 *General Context*

The Dominican Republic (DR) is a developing country situated in the Caribbean region between Cuba and Puerto Rico, sharing the same island as Haiti. A general portrait of the country can be summarized in the following data from Central Intelligence Agency (CIA 2016):

1. Official language: Spanish
2. Population: as of July 2016, 10.6 million people
3. Religion: Roman Catholic (95%), Others (5%)

[14]The 2016 E-government survey (United Nations Department of Economic and Social Affairs, 2016) concluded the following regarding the use of Open Government Data: "In an effort to make public institutions more inclusive, effective, accountable and transparent, as called for in the 2030 Agenda for Sustainable Development, many governments across the globe are opening up their data for public information and scrutiny. Making data available online for free also allows the public—and various civil society organizations—to reuse and remix them for any purpose. This can potentially lead to innovation and new or improved services, new understanding and ideas. It can also raise awareness of governments' actions to realize all the SDGs, thus allowing people to keep track and contribute to those efforts. Overall, in 2016, 128 out of 193 UN Member States provide datasets on government spending in machine readable formats. The remaining 65 have no such information online (p. 2)".

4. Urbanization: 79% of total population
5. Literacy: 91.8%
6. GDP per capita: USD $15,000; 108th place in the world
7. Labor force by occupation: Agriculture (14.4%), Industry (20.8%), Services (64.7%)
8. Economic overview: The country suffers from marked income inequality; the poorest half of the population receives less than one-fifth of GDP, while the richest 10% enjoys nearly 40% of GDP. The economy is highly dependent upon the United States of America (USA), the destination for approximately half of exports. Remittances from the USA amount to about 7% of GDP, equivalent to about a third of exports and two-thirds of tourism receipts.

The DR, like other Caribbean economies, shares many of the characteristics of small States: open and vulnerable economies, narrow resource bases, limited diversity in production, exports concentrated on a few products, thin markets and high transportation costs (CEPAL 2013a, p. 6).

Regarding ICT infrastructure and development, the DR has an NRI of 3.6, occupying the 98th spot out of 139 countries (Baller et al. 2016) and an IDI of 4.30, occupying the 104th spot out of 175 nations (International Telecommunications Union 2016). As has been the tendency in other developing countries, the DR has seen an important increase in mobile phone internet accounts, as we can see in the following graphic (Fig. 6).

At the same time, the percentage of homes with internet service has gone from 3.1% in 2005 to 23.6% in 2015; and the percentage of homes with computers has gone from 8.7% in 2005 to 30% in 2015 (Instituto Dominicano de Telecomunicaciones 2016). Nonetheless, there is still an important divide among the poorest and the richest groups of citizens regarding households with computers, as we can see in this chart (Quintil I being the poorest and Quintil V the richest):

As we can see, there is an important Digital divide between the poorest and richest citizens of this country. In the case of the Latin America and Caribbean Region (LACR), the CEPAL concluded the following regarding the Digital divide (Fig. 7):

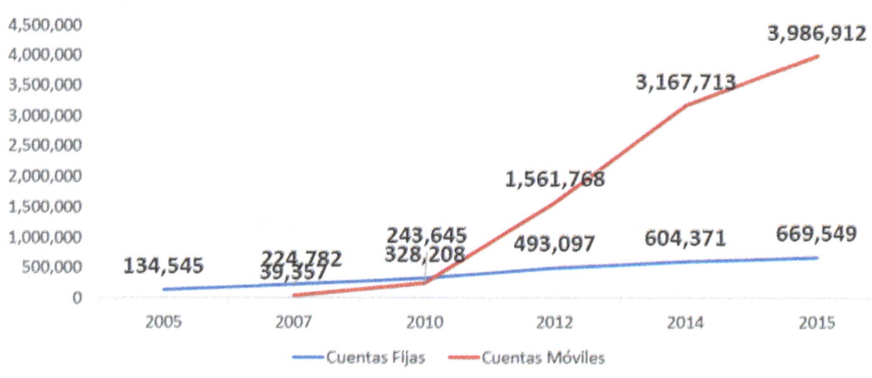

Fig. 6 Instituto Dominicano de Telecomunicaciones (2016, p. 3)

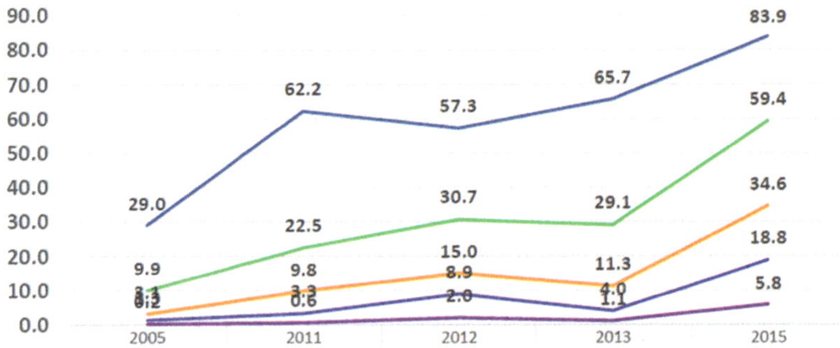

	2005	2011	2012	2013	2015
Quintil I (más pobre)	0.2	0.6	2.0	1.1	5.8
Quintil II	1.3	3.3	8.9	4.0	18.8
Quintil III	3.1	9.8	15.0	11.3	34.6
Quintil IV	9.9	22.5	30.7	29.1	59.4
Quintil V (más rico)	29.0	62.2	57.3	65.7	83.9

Fig. 7 Instituto Dominicano de Telecomunicaciones (2016, p. 13)

> The digital divide in terms of income, localization, ethnicity and gender, both among the countries in the region and between the region and the developed economies, has not closed significantly, especially with regard to access to high-speed mobile broadband. The distance is even greater if we consider not only these technologies, but also new platforms associated with mobility, cloud computing, social networks and the big data analytics for decision-making (CEPAL 2013b, p. 5).

Under those circumstances, we shall now briefly explore the digital strategies taken on by the DR in order to reduce its Digital divide.

4.2 Digital Strategies in the Dominican Republic

The process of digitization and structural change is not spontaneous, but rather is generated in institutional contexts that are conducive to innovation and that promote the development of ICT sectors, providing incentives to rapidly diffuse new applications to non-ICT sectors of the economy (CEPAL 2013b, p. 99). Thereupon, The Economist Intelligence Unit (2015, p. 37) has stated that the challenge for Caribbean economies is to create domestic industries that generate jobs and boost economic growth, relying less on government as a source of employment and growth.

As we have stated, ICTs can dramatically help domestic industries in creating jobs and boosting economic growth through the mechanisms of the digital and data revolutions. In the case of the DR, the current situation from a strategic point of view (as of October 2016) is that there are two digital strategies waiting to be developed by government authorities: *República Digital* (Digital Republic) and *Agenda Digital República Dominicana 2016–2020* (Digital Agenda for the Dominican Republic 2016–2020).

Let's start our analysis by examining the *República Digital* Strategy (RDS). This is a project that stems from the presidential campaign of Danilo Medina,[15] who was reelected for a second term on May 2016. Up until the moment we were writing this article, there has not been a full disclosure of this strategy, other than the contents of the campaign material, which amounts to less than three pages. Nonetheless, the RDS is comprised of the following objectives and targets (Instituto Dominicano de Telecomunicaciones 2016, p. 17):

(1) Education and technology:

   1. Computers for all students and teachers of grade school.
   2. Digital abilities courses for students and teachers of grade school.

(2) Universal Broadband access:

   1. Construction of a national broadband network for all provinces.

(3) Digital productivity:

   1. Ample incentives, education and support for SMEs in terms of internet, ICTs, e-commerce, etc.

(4) Digital and transparent government:

   1. Increase of e-government services.
   2. Increase of the transparency of public institutions.

On the other hand, we have the *Agenda Digital República Dominicana 2016–2020* Strategy (Agenda), authored by the National Commission for the Knowledge and Information Society (*Comisión Nacional para la Sociedad de la Información y el Conocimiento*), created in 2005 and presided by the Dominican Telecommunications Institute (Indotel), which is the most specialized public institution concerning ICTs and Technology in the country. The Agenda, now in its fourth version,[16] is supposed to be the national strategic plan that establishes the guidelines in order to enable social and economic development through ICTs (Comisión Nacional para la Sociedad de la Información y el Conocimiento 2015, p. 3); and its objectives match those established by the Dominican National Development Strategy for 2030.[17] The document has 17 specific objectives, 39

[15]The website for the campaign can be accessed at: www.holard.do.

[16]The current version is a preliminary one, awaiting input and comments from the public via a consultation at the time we wrote this article, that would end in September 2016. For more information visit: http://www.cnsic.org.do/index.php/agenda-digital.

[17]This strategy was approved by the national congress in January 2012 and contains the long term vision, as well as the targets to make the Dominican Republic "a prosperous nation, where people live with dignity, upholding ethical values in a social and democratic state settled upon a participative democracy that promotes equity, equal opportunity, social justice and manages and takes advantage of its resources in an innovative and sustainable way, inserting itself correctly in the global economy" (Ministerio de Economía Planificación y Desarrollo 2016, p. 3). Regarding ICTs, we can conclude that the term "technology" is present throughout the whole document, including a

actions and 115 initiatives. Its five strategic pillars are (Comisión Nacional para la Sociedad de la Información y el Conocimiento 2015, p. 6):

1. Infrastructure and access: to facilitate broadband internet access to the population in an affordable way.
2. E-government and digital services: to have a participative and transactional e-government, ensuring confidence, security and privacy.
3. Capacity building: to develop digital competencies in the population so that the country can have the necessary human capital to sustain a digital economy.
4. Innovative and productive development: to increase the levels national competiveness through ICTs.
5. Stimulating environment: to build a stimulation environment that facilitates the advancement of the Knowledge and Information Society in the country.

As we can see, both strategies have very similar general objectives, but differ in one main aspect: while the Agenda has been the work of various governmental and private organizations throughout the past ten years, the RDS has been the work of a close group of advisors commissioned by the current President.[18] Consequently, we strongly believe that the Agenda will be substituted by the RDS in the months to come, due to the importance given by the President in the past election campaign.

Although the full details of the RDS are not available at this time, we can safely say that it will incorporate some of the most pressing ICT development issues in the region. Accordingly, an expert survey on priorities for the information society in the Caribbean commissioned by the Economic Commission for Latin America and the Caribbean (Crane Williams 2016) concluded that the most voted strategic goals for the region where: to "Promote teacher training in the use of ICTs in the classroom" (RDS objective i) and to "Reduce the cost of broadband services" (RDS objective ii).

If the DR (or any other developing country in Latin America) wishes to implement a successful digital revolution, one of the pillars must be the availability of broadband services at a reasonable price according to the local economy. Consequently, the OECD and the IDB (2016, p. 16) have recommended the following good practices in order to increase broadband access in Latin America and the Caribbean:

1. Digital strategies and national broadband plans should seek to increase broadband access and usage by using a whole-of-government and multi-stakeholder approach.[19]

(Footnote 17 continued)

series of objectives and targets associated with them, most notably "Specific Objective 3.3.5" which looks to "achieve a universal and productive access to ICTs" (Ministerio de Economía Planificación y Desarrollo 2016, p. 59).

[18]We have come to this conclusion through the statements of friends who work in various public institutions in the DR and have required us to not unveil their identity.

[19]In 2015, the *Instituto Dominicano de Telecomunicaciones* announced that it would implement a national fiber optic program in order to provide broadband access to all the municipalities in the country. For more information see: http://indotel.gob.do/proyectos/red-nacional-de-fibra-optica-2/.

2. A stable and predictable regulatory framework is necessary to cultivate long-term investment in broadband infrastructure. Sound regulations can help expand infrastructure expansion by lowering the costs of deployment.
3. Broadband should be made increasingly accessible and affordable to disadvantaged groups and people living in rural and remote areas.
4. Regional co-operation arrangements, sharing of regulatory experiences,[20] deployment of regional connectivity infrastructures, cross-border data flows and lowering the prices of international connectivity and roaming should be encouraged.
5. Broadband services should be made available in schools, health care centers and other places of public access,[21] along with the promotion of a skills system geared to the digital economy.
6. Facilitating ICT adoption by businesses, creating digital content accessible to local populations, and the promotion of digital entrepreneurship can all increase demand and improve services.
7. Digital governments should be actively promoted in the LAC region to allow for smarter organization of cities and to help governments become more efficient, effective, open, transparent and accountable.[22]

On the business part, we believe that one of the key areas for the RDS will have to do with the creation of a true Dominican Digital economy via the transformation of current Dominican companies into digital enterprises. According to the World Economic Forum and Accenture (2016), becoming a digital enterprise requires far more profound changes than merely investing in the latest digital technologies. Thus, companies will need to search for new business models, fundamentally rethink their operating models and revamp how they attract and foster digital talent.

The RDS objective iii (Digital productivity) takes into account some of these recommendations through the following measures (Instituto Dominicano de Telecomunicaciones 2016):

1. Incentive programs that foster digital and technological entrepreneurship.

---

[20]The *Instituto Dominicano de Telecomunicaciones* regularly participates in all kinds of regional and international cooperation and regulatory events. One recent event was promoted by the Organization of American States (OAS) and involved the signing of an agreement looking to foster public-private partnerships towards the development of ICT infrastructure in the region. For more information see: http://indotel.gob.do/oea-celebrara-en-rd-foro-sobre-uso-tic-para-erradicar-la-pobreza-en-las-americas/.

[21]One of the authors of this article worked for several years in a government Project aimed at providing Access to Internet through the construction of government funded Technological Community Centers (Centros Tecnológicos Comunitarios). For more information see: http://www.ctc.edu.do/.

[22]The *Oficina Presidencial de Tecnologías de la Información y Comunicación* (ICT Presidential Office) is the public institution in charge of developing and supervising the Dominican e-government strategy. They have even developed an index that measures the e-government adoption rate of the main public institutions. For more information see: http://sisticge.dominicana.gob.do/ and http://www.optic.gob.do/.

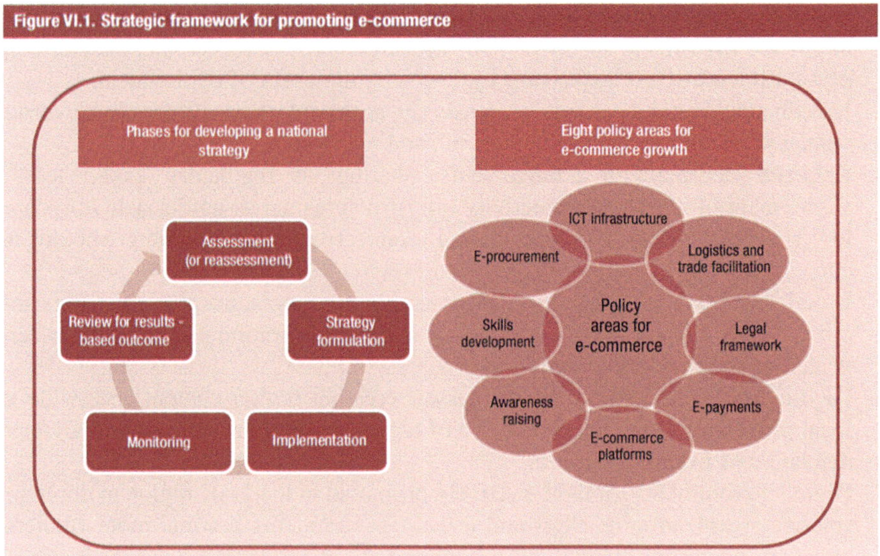

Fig. 8 United Nations Conference on Trade and Development (2015, p. 84)

2. The increase of international scholarships.
3. Advisory programs for small and medium companies.

As we can see in the following graphic, the RDS would fall into the phase of "Strategy formulation" in relation to the Strategic framework for promoting e-commerce of the United Nations Conference on Trade and Development (2015) (Fig. 8).

In this framework, we can see that some of the principle policy areas that foster e-commerce growth have a direct connection with the data revolution mechanisms we have described. For example, the emergence of Big data analysis can help raise awareness towards social issues affecting a country (Bellagio Big Data Workshop Participants 2014) or can facilitate trade and logistics within a country or a region (Federal Trade Commission 2016).

5 Conclusion

In conclusion, we strongly believe that the digital revolution has the potential to accelerate the social and economic transformations needed in developing countries in order to improve their human development performance. On the infrastructure side, broadband access must continue to be a priority in every digital revolution strategy, given the new Digital divide gaps that have emerged in the past ten years.

On the social side, Big data for development can be one of the premier mechanisms that drives the data revolution within the bigger digital revolution picture.

The Dominican Republic is one of the developing countries which seems to be on the right track from a strategic and theoretical perspective. If it can overcome certain political and organizational issues, its citizens are on track to reap important social and economic improvements in their everyday livelihoods.

References

Baller S, Dutta S, Lanvin B (eds) (2016) The global information technology report 2016. World economic forum. Retrieved from https://www.weforum.org/reports/the-global-information-technology-report-2016/

Bellagio Big Data Workshop Participants (2014) Big data and positive social change in the developing world: a white paper for practitioners and researchers. Oxford Internet Institute, Oxford. Retrieved from http://www.rockefellerfoundation.org/uploads/files/c220f1f3-2e9a-4fc6-be6c-45d42849b897-big-data-and.pdf

Berger-Walliser G, Shrivastava P (2015) Beyond compliance: sustainable development, business, and proactive law. Georgetown J Int Law 46(2):417–474

Centre for International Governance Innovation, The Royal Institute for International Affairs (2016) Global commission on internet governance. Chatam House

CEPAL (2013a). Caribbean regional synthesis report. Third international conference on small Island developing states. Retrieved from https://sustainabledevelopment.un.org/content/documents/5164250SynthesisReportFinal20August2013.pdf

CEPAL (2013b) The digital economy for structural change and equality. Retrieved from http://repositorio.cepal.org/bitstream/handle/11362/35954/1/S2013350_en.pdf

CIA (2016) CIA world factbook. Retrieved from https://www.cia.gov/library/publications/the-world-factbook/geos/dr.html?

Comisión Nacional para la Sociedad de la Información y el Conocimiento (2015) Agenda Digital República Dominicana 2016–2020. Versión preliminar. Retrieved from http://www.cnsic.org.do/images/docs/Agenda/Versión-Preliminar-Agenda-Digital-R-D-2016-2020-para-Consulta-Pública.pdf

Crane Williams R (2016) Expert survey on priorities for the information society in the Caribbean. Economic Commission for Latin America and the Caribbean (CEPAL). Retrieved from http://repositorio.cepal.org/bitstream/handle/11362/40255/1/LCCARL491_en.pdf

Economist (2014) The rise of the cheap smartphone. Retrieved from http://www.economist.com/news/business/21600134-smartphones-reach-masses-host-vendors-are-eager-serve-them-rise-cheap

Economist (2015) Planet of the phones. Retrieved from http://www.economist.com/news/leaders/21645180-smartphone-ubiquitous-addictive-and-transformative-planet-phones

Economist Intelligence Unit (2015) Private sector development in the Caribbean: a regional overview. Retrieved from http://www.eiu.com/public/topical_report.aspx?campaignid=CompeteCaribbean0515

Federal Trade Commission (2016) Big data. A tool for inclusion or exclusion? Retrieved from https://www.ftc.gov/system/files/documents/reports/big-data-tool-inclusion-or-exclusion-understanding-issues/160106big-data-rpt.pdf

Heeks R (2014a) From the MDGs to the post-2015 agenda: analysing changing development priorities. Development informatics. Working paper series no. 56

Heeks R (2014b) ICTs and poverty eradication: comparing economic, livelihoods and capabilities models. Development informatics. Working paper series no. 58

High-Level Panel of Eminent Persons on the Post-2015 Development Agenda (2013) A new global partnership: eradicate poverty and transform economies through sustainable development. Retrieved from http://www.post2015hlp.org/wp-content/uploads/2013/05/UN-Report.pdf

Instituto Dominicano de Telecomunicaciones (2016) República Digital: Plan de República Dominicana para la inclusión en el uso de las TIC's. Retrieved from https://www.itu.int/en/ITU-D/Regional-Presence/Americas/Documents/EVENTS/2016/15544-BR/1-2.pdf

International Institute for Sustainable Development (2016) Sustainable development. Retrieved from http://www.iisd.org/topic/sustainable-development

International Telecommunications Union (2016) ICT development index 2016. Retrieved from http://www.itu.int/net4/ITU-D/idi/2016/#idi2016rank-tab

Ministerio de Economía Planificación y Desarrollo de la República Dominicana (2016) Estrategia nacional de desarrollo 2030. Retrieved from http://economia.gob.do/mepyd/wp-content/uploads/archivos/transparencia/base-legal/end_2030.pdf

OECD (2016a) Trends in shaping education 2016. Retrieved from http://www.oecd-ilibrary.org/education/trends-shaping-education_22187049

OECD (2016b) Development co-operation report 2016. Retrieved from http://www.oecd-ilibrary.org/development/development-co-operation-report-2016_dcr-2016-en

OECD, Interamerican Development Bank (2016) Broadband policies for Latin America and the Caribbean. Retrieved from http://www.oecd-ilibrary.org/science-and-technology/broadband-policies-for-latin-america-and-the-caribbean_9789264251823-en

Sen A (1985) Well-being, agency and freedom. J Philoso LXXXII 4:169–221

UN (2006) Millennium project. About MDGs. Retrieved from http://www.unmillenniumproject.org/goals/

UN (2015) The millennium development goals report 2015. Retrieved from http://www.un.org/millenniumgoals/2015_MDG_Report/pdf/MDG2015rev(July1).pdf

UNESC (2014) Information and communications technologies for inclusive and economic development. Report of the secretary-general. E/CN.16/2014/3

UNESC (2015) Digital development. Report of the secretary-general. E/CN.16/2015/2

UNESC (2016) Multi-stakeholder forum on science, technology and innovation for the Sustainable Development Goals: summary by the Co-Chairs. E/HLPF/2016/6. Retrieved from http://www.un.org/ga/search/view_doc.asp?symbol=E/HLPF/2016/6&Lang=E

UNESCO (2014) UNESCO education strategy 2014–2021. UNESCO. Retrieved from http://unesdoc.unesco.org/images/0023/002312/231288e.pdf

UNGA (1986) Declaration on the right to development. Retrieved from http://www.ohchr.org/Documents/Issues/Development/DeclarationRightDevelopment_en.pdf

UNGA (2015) Transforming our world: the 2030 Agenda for Sustainable Development. A/RES/70/1. Retrieved from http://www.un.org/ga/search/view_doc.asp?symbol=A/RES/70/1&Lang=E

United Nations Conference on Trade and Development (2015) Information economy report 2015. Unlocking the potential of E-commerce for developing countries. Retrieved from http://unctad.org/en/PublicationsLibrary/ier2015_en.pdf

United Nations Department of Economic and Social Affairs (2016) United Nations E-government survey 2016. United Nations. Retrieved from https://publicadministration.un.org/egovkb/Portals/egovkb/Documents/un/2016-Survey/ExecutiveSummary.pdf

United Nations Development Programme (2015) What is human development? Retrieved from http://hdr.undp.org/en/content/what-human-development

United Nations Development Programme (2016) Human development index. Retrieved from http://hdr.undp.org/en/content/human-development-index-hdi

United Nations Secretary-General (2014) A world that counts. Mobilising the data revolution. Retrieved from http://www.undatarevolution.org/wp-content/uploads/2014/12/A-World-That-Counts2.pdf

World Economic Forum, Accenture (2016) Digital transformation of industries. Enero 2016: World Economic Forum White Paper

Author Biographies

Luis A. García-Segura Professor of Law and Coordinator of the Nebrija-Santander Chair on Risk and Conflict Management at Nebrija University, Madrid, Spain. Earned his Master's Degree in Business Law in 2010 and his Ph.D. in Law in 2015. Expert in Business Law, professional services industry, cybersecurity and cyber defense. He worked for the Dominican government from 2007 to 2009 in an educational project to help reduce the Digital divide gap. He is also a practicing lawyer, admitted to the Madrid Bar Association (ICAM) and a faculty member of the American Bar Association (ABA).

Juan Cayón Peña Rector at Nebrija University, Madrid, Spain. Expert in Natural law, cybersecurity, cyber defense, privacy, strategy and intelligence, has numerous publications in various legal related fields. He is also a practicing lawyer, admitted to the Madrid Bar Association (ICAM). Dr. Cajón is a Member of the Spanish Royal Academy of Jurisprudence and Legislation and a Fellow of the World Academy of Art and Science.

Business Strategy in the Digital Age. Digital Transformation, Disruption and Cybersecurity

Manuel Gago-Areces

Abstract We are at the beginning of a new industrial revolution, the fourth industrial revolution, which means the total connectivity with support of mobility and a revolution in resources. This allows real-time decision making, in the lines where the action is performed, to gain efficiency, speed and adjustment to local conditions and a centralized control with structures in which the power of approximation is centralized, forming ecosystems with a dynamic orientated to project actions in real time with a shared and accepted criteria. The new IT and telecommunication technologies have made possible the MANAGEMENT INFORMATION SYSTEM in REAL TIME, supporting the network of connections that identifies who are the participants and they do, decide and access the information only in an authorized manner thanks to Cybersecurity models. This concept has more obstacles than technological ones; those are the mindset of the participants and especially deciders, due to their lack of knowledge or experience to understand an interconnected globally with disruption. The speed will change the nature of the business when the velocity of change be enough, the nature of the business will be completely changed. These changes will happen due to a single factor: The digital information flow and its availability in real time. Digital transformation means disruption, not only to optimize process, but also to focus on new solutions, products and resources, giving a globally net with worldwide connectivity. But not all are good news, the digital transformation also means risks and dangers, (remembering Escila and Caribdis, being between two dangers, avoiding one could be in the grasp of the other). One wrong vision could change strengths into weaknesses and in our connected era the reversion works with such devastating velocity that a Company could be led from a quiet equilibrium to the chaos instantly. The attractive idea than connection means liberation is not only a mistake

Submitted: 8-12-2016; Accepted: 29-12-2016.

M. Gago-Areces (✉)
Neo Soluciones Informáticas, S.L., Madrid, Spain
e-mail: mgago@neo-si.com

J.M. Ramírez and L.A. García-Segura (eds.), *Cyberspace*, Advanced Sciences and Technologies for Security Applications, DOI 10.1007/978-3-319-54975-0_5

because it could mean also get trapped to a strong dynamic net with many difficulties to get liberated from. At the same time who is not connected has not value at all. In order to avoid digital risks it is necessary to take very complex measures due to the complexity of the systems, such as: Designing a Strategic Information System with control and firewalls under responsibility of a Chief Technical Officer (CTO), protecting Data Bases, blocking data security encrypted, fragmentation and other mechanisms to assure impossible to be robbed or manipulated, with "Web Doorkeepers" watching the net and the interconnectivity of all the ecosystems where the Company is and to implant Internet of Things (IoT) avoiding security leaks, with controls in process, input and output to avoid manipulations with the trust of all the people involved. But trust is what every company needs to be in a competitive environment, knowing that nobody can reject progress. Developing one trustworthy environment inside and outside the Company giving confidence to owners, employees, customers, suppliers, and financial and official institutions. At the same time act firmly (dare to know) and never interrupt the progress.

Keywords Business · Digital transformation · Cybersecurity · Network

1 Progress of the New Technologies, Networks and Consequences

In the next ten years the companies will change more than in the last fifty years. If the decade of the eighties of the last century was the decade of quality and the nineties of reengineering process, the first decade of the present century has been that of the internet, for today we live in the speed connectivity, modifying constraints of time and space. We are at the beginning of a new industrial revolution and like the previous it ones also brings with it new organizational structures and management practices.

The first revolution had its reference in the steam engine in 1712, which changed the sweat by the steam and traditional forms of work by overcoming a world of craftsmen's guilds.

The second one with electricity in 1831, creates factories, centers activity groups, giving way to the first assembly lines and large public corporations and strategic management (the man of the organization).

The third revolution occurred with internet and especially, in the last decade of the last century making it possible for the dynamics of processes integrating information and databases with communications. Thus, the previous stages were eclipsed, making the reengineering process much easier, changing the way customers and consumers buy, and incorporating speed as the engine of transformation.

The fourth industrial revolution, which we are experiencing nowadays, is the beginning of total connectivity with the support of the mobility. It allows us to make displaced decisions in real time, in the lines where the action is performed, to gain efficiency, speed and adjustment to local conditions and a centralized control with structures in which the centralization/decentralization approach is maintained. We could call it a networking organization, forming ecosystems with dynamic orientated to the action and a strategic approach based on the opportunity supply of information of the performed operations and with this knowledge to project actions in real time with a shared and accepted criteria.

The new IT and telecommunication technologies, particularly online systems, have facilitated to information systems to change not only production processes, in commercial systems or in administration, but also management process, giving response to the needs that the time factor has introduced in the management. Thus, they allow a management oriented to the action, with interactive support of actions, thus configuring a global direction system in real time, to join all intellectual resources together to solve a serious problem by providing the measure of itself according to how quickly a difficulty is detected and reacted to (Welch 2002; Immelt 2004).[1]

The time factor is the basic element of convergence in management systems nowadays (Quality, Labor Equipment, Lean, 6sigma, Redesign Processes, Change Management…) being the capacity of response in time (in the same as in controlling the situation at the moment) the key to achieve a competitive position. Let fly the news and better with anticipation if they are bad; but it is useless to be alert without the tools that can channel this news by the organization and allow to act quickly (Gates 1999). IT and specially the "mobility" allows us to overcome space/time barriers, uniforms procedures, to interpret and facilitate decisions where the activity is carried out (displaced), instantly recording the facts and their consequences, and ensuring an efficient centralized control.

It is also necessary open communications that support the connections network that identifies safely during all the time who are the participants and that these ones act, decide and access to the information only in an authorized manner with Cybersecurity.

This concept, which applies to all types of structures, whether individuals, companies, civil organizations or public administrations and local, national or supra-national levels, must overcome more important obstacles than technological

[1]Welch (2002), former CEO of General Electrics (GE) considered "the organization's ability to learn and translate quickly what it was learned into action, as the ultimate competitive advantage" and Immelt (2004), the current President of GE, proposed: "break with the historical organization of processes, create a culture of simplification fewer rules, fewer processes, fewer decision points, adapt lean tools with a Silicon Valley approach and install a digital thread that covers the whole structure of the organization, input, process and output and have everything in time, fast work, with commercial intensity throughout the company, disconnect with the annual, and carry out closer follow-ups, with continuous reviews and covering a 360-degree horizon".

ones, as could be the participant minds and especially those with management responsibility, in other words, those who decide, because:

1. some of them lack the knowledge or lack of experience to understand an interconnected globally, are cyber-ignorant;
2. others, due to the inability to adapt to the concept of instantaneity breaking the time/space relationship, which disruptively changes processes, eliminates functions and tasks and enables permanent presence even when it is virtual.

The new concept has allowed the creation of new organizations, with extraordinary success, incorporating the instantaneity of local input with decision-making in the traditional centralized model.[2]

It also allows a flexible management model, but in order to reduce or eliminate risks, it is necessary to improve control systems against external interferences and, above all, to ensure the confidentiality of information and the correct execution of the processes. Cybersecurity systems that allow the availability and protection of information and decisions in a timely manner, that the relevant information flows from the bottom up and from displaced points, remote from the central area and vice versa, breaking space/time barriers and in such a way that it is almost impossible to suffer any consequence of unauthorized access or tampering both from the outside and from within the system.

In order to apply the disruptive potential that new technologies have, it is essential to carry them out in a fast way to achieve significant changes in income or savings, avoiding the market can opt for other products and services offered by competitors. If there is not an advance from the expectations of the market, the company will be expelled.

The success of the new model is shown when new attributes in the processes, resources, products or services appear and at the same time, in a vertiginously and continuous course, the moment to search for a new substitution is indicated.

The combination of experiences in companies, universities, or research projects, ensure that we have the tools to achieve full impact in setting strategies. The progress has been the result of many efforts dedicated to research, development of analytical techniques, and experimentation in models of specific business situations as well as management programs given by prestigious business schools, especially by Silicon Valley.

The enterprise system becomes recognized not only as a set of separate functions with a common purpose, but as a system in which the flows of information, materials, labor, capital goods and money establish the forces that determine trends

[2]Examples such as Amazon, Zara, or Uber have changed the retail distribution, the "*instant mode*" model, transportation services, with more control over the total process, from design, procurement, manufacturing, distribution and marketing, or vertically integrating decisions and putting all the emphasis on the speed of access to the market and providing the customer with instant communication of the service. Pioneering companies (like GE, ZARA, CISCO, IBM and others) have been the introducers, improving their position in their markets, their industrial sector and the national economy. Competitive pressures have made other followers look for the same advantage.

towards the growth, the fluctuations or the possible fall and bankruptcy. Not being the idea of a movement and the function relationship what matter, but the constant change of these functions and their relations as dynamic activities. The new vision that implies the incorporation of new technologies, digital transformation and mobility into the Business management system, as a complex concept, will take time (as in any important social change), in order to develop the ideas and for the incorporation of a new generation of managers.[3]

The speed with which the nature of companies will change, the speed with which the transactions will take place, how to access to the information, and so on, will change the lifestyle of the consumers and the expectations of companies. The quality and processes improvements will be introduced much more quickly, and when connectivity and speed are sufficient, the very nature of the enterprise will be transformed. Even though we have been in the Information Age for more than thirty years, only very recently it has been reduced what was the main information support for the companies, the paper, but still most of the processes are automation of the classic processes.

These changes will happen due to a single factor: the flow of digital information and its availability in real time (Fig. 1).

2 Digital Transformation and Disruption

Digital transformation implies disruption, and not only process optimization, which in many cases eliminate them in combination with components and products, but to focus on new solutions, products and resources as a consequence of the digitization and connectivity.

Frequently those who lead takes decision based on the absence of up-to-date information, in spite of having instruments of information online. Even somebody who has invested heavily in information technology can get poorer results than expected, mainly for cultural reasons, or for an incomplete understanding of the possibilities, combined with a lack of vision of the disruptive potential that technology offers.

[3]Many companies are already developing the necessary skills for this with engineering departments and operational researches, new simulation techniques, development of feedback systems and control and analysis of decision processes. For them, the management research is the process of a long-term planning for the improvement of the company. But these companies are the exception and their implementation has not yet been completely appreciated. At the same time, many companies fail because of mismanagement rather than because of a bad product or service or poor engineering. Very few companies use digital technology to create new processes that radically improve their operation, changing or eliminating them, and extracting the full performance of their workforce and their capacity for knowledge and speeding up new ways of competing, in a disruptive way, in an already consolidated globalized market.

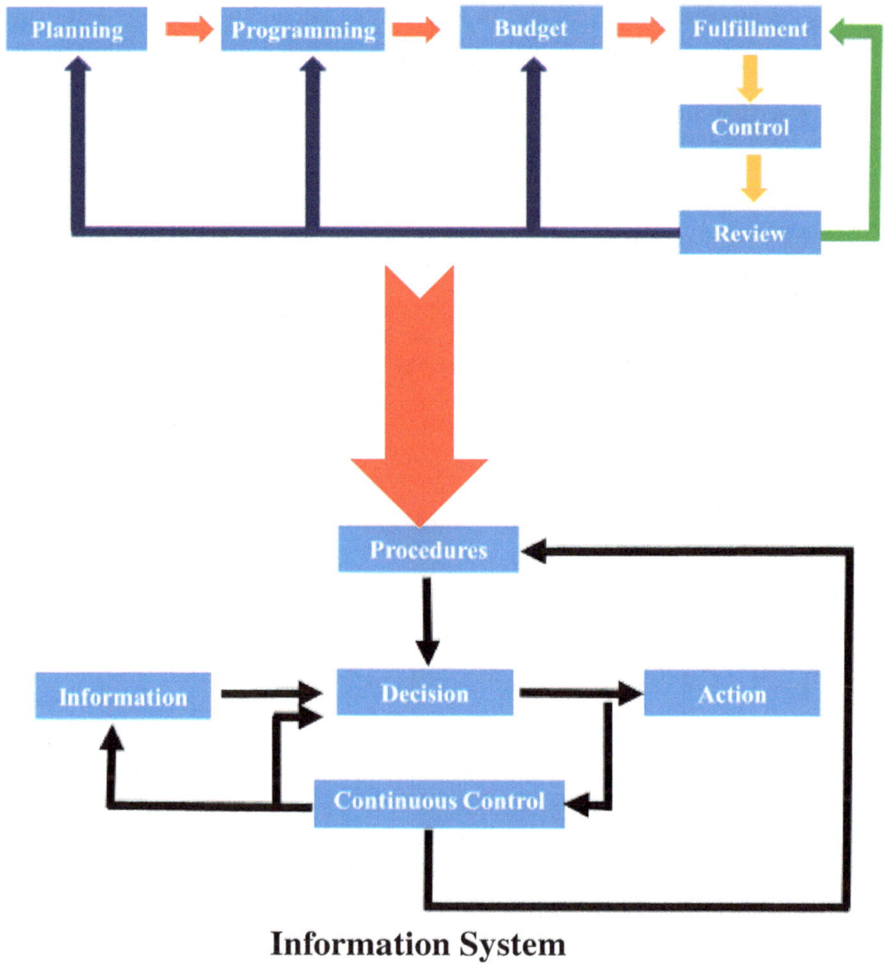

Fig. 1 Cybernetic model

Today, digital technology puts in our hands the means to obtain information, share it, decide and act on it in many ways, expressing digitally all kinds of information, text, data, graphics, voice and sound, images and video.

At the same time, are being created a new generation of smart digital devices, such as smartphones, handhelds, smart cards and others. Through them the use of digital information will be widespread and with the improvements of communication and Internet networks they will have global connectivity.

The Network creates a new universal space of sharing information, and assuming the immediate and spontaneous nature of television and telephone, combined with the ability to offer information to a group of people with common interests.

New forms emerge for the physical and the logical to share communications, altering not only the processes but also the style of management, work and, of course, the lifestyle, which needs to adapt to the new digital reality.

The substitution of documents (paperwork) for digital processes has considerably eliminated expenses and delays, providing work-teams with intelligent digital devices that act at the speed of an individual who possesses the knowledge of the whole team, taking advantage of the mental capacity of all their components.

With real-time information we are able to react with more agility to the problems and opportunities, configuring the Information Decision System as the human nervous system, staying alert to the important things and providing selected information to evaluate options and take decisions.

The companies will have to face the performance of a process of transformation with the following aspects:

1. Identify how digital transformation can affect the business.
2. Lead change. Involving the top managers of the company to support and drive the process, eliminating uncertainties, internal frictions and apathy for being in consolidated comfort zones. It is not enough to appoint a CTO (Chief Technology Officer), responsible for this change, it also implies commitment, alignment and changes throughout the organization.
3. Model as much as possible and test with MVP (minimum value processes) criteria before a global execution.
4. Communicate the change. Confirming the culture of the company, make a fluid communication to the people involved on the objectives of the transformation, the advances and results that are being achieved, the failures, the analysis of possible improvements in a dossier of good practices.
5. Passion for innovation: attitude with orientation, or rather obsession, with the process of change and especially with the technology to improve the position before the clients by making them participate.
6. Effective management. The digital transformation introduces a new way of managing the projects, supported by continuous improvement, breaking with the bureaucracy, incorporating disruption. Experimenting on MVP models to facilitate test, minimizing risks.
7. Control and monitoring. Define control metrics and key indicators, quantitative objectives, KPI (Key Performance Indicators) that effectively measure the level of performance of processes, their performance and the degree of achievement of the objectives set.
8. Cybersecurity. Establishing controls over the network, access to information, databases and decision processes. CTO and Doorkeeper.
9. Alliances and outsourcing. The time factor is key so it will be necessary to have outreach supports that help reduce implementation without risk of competition, saving learning time and incorporating professionals with the specific experience necessary to achieve the best efficiency.

In order to be disruptive, digital transformation should be based on five principles:

1. Ability to identify opportunities for the substitution of components, resources, especially the scarce ones.
2. Eliminate/minimize waste, from procurement, production to the end user.
3. Increase circularity by updating, reusing, or recycling materials and components.
4. Optimization, efficiency, convenience, security and reliability.
5. Eliminate products, services and processes that can be done outside the physical world, within the virtual environment.

A Digital Strategy means a combination of physical equipment and infrastructures with software, distinguishing the Information System from the simple computerized network, for the ability of collaboration, precision and instantaneity of decisions and answers.

Nonetheless, digital transformation also involves dangers and risks, which reminds us of the expression "between Scylla and Charybdis" meaning to be between two dangers so that getting away from one can cause falling into the other. Disappear by inactivity or die by too much anticipation. "Change or die". A deep commitment to a wrong vision can turn strengths into weaknesses and in our connected era, this kind of reversion acts with a particularly devastating speed, being able to move from equilibrium to chaos immediately. In the next few years, I think that we will see some leading business giants disappear as has was the case with the dinosaurs.

Just as the products and services can have three characteristics: good, fast and cheap, but only two of the three at a time, in the networks, the systems can be: fast, open or secure, but only two at a time.

The attractive idea that connection means liberation is a mistake because it could involve getting trapped in a strong dynamic net with many difficulties to get liberated from. At the same time who is not connected has no value at all. The attractive idea than connection means liberation is not only a mistake because it could mean also get trapped to a strong dynamic net with many difficulties to get liberated from. At the same time who is not connected has not value at all.

The business risks of a company are threats that a fact or action may adversely affect the ability of the organization to meet its objectives and strategies successfully. Digital opens us to new risks that require a new way to cover them and be part of a specific strategy of risk management. In order to manage organization risks, managers need to focus on effective processes and rely on experts with relevant knowledge (Fig. 2).

The more devices are connected to the network, the more power the central hub gathers. In a connected system, power is defined by both deep concentration and mass distribution.

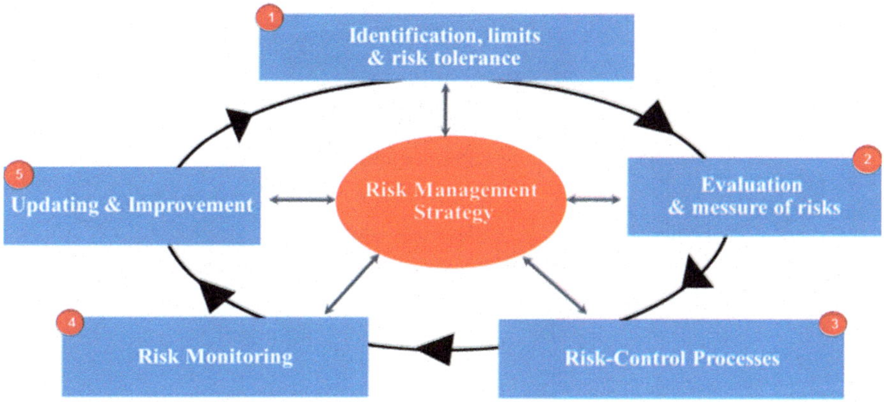

Fig. 2 Principles of digital transformation

3 Convergence in the Changes and Impact on Business

Companies face major challenge as a result of the new environment and the profound current changes:

1. Economical:
   - Continuous Process of Globalization.
   - Debt Crisis.
   - Climate change.
   - Energy and resources crisis.
   - Questioning of Economic Liberalization.
   - Dollar-Euro-Yuan-Yen—New Currencies?
2. Technological:
   - Disruptive digital transformation.
   - New technological developments "shared" all over the world.
   - Costs of technology falling significantly.
   - Internet-Mobility Connectivity.
   - Cybersecurity.
3. Political-social:
   - Migratory Movements and Expatriates.
   - Populism.
   - Nationalisms.
   - Demographic problems.
   - Equality and gender.
   - Wars (Fig. 3).

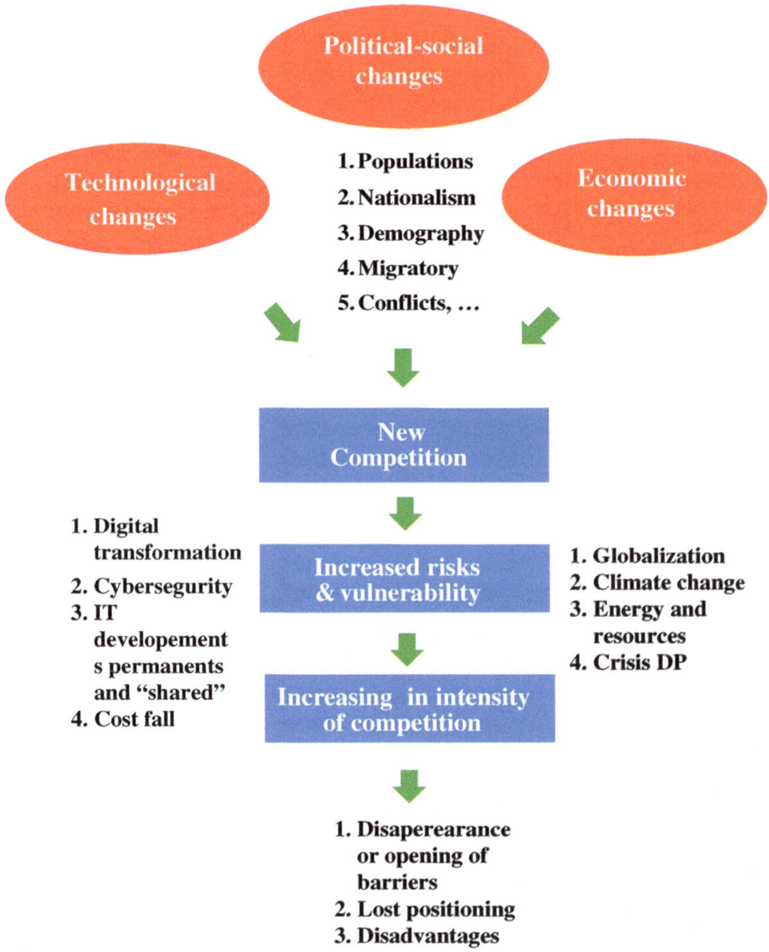

Fig. 3 Major challenges and changes of companies

All these factors have a common effect: the continuous circularity with elimination of barriers, facilitating the increase of competition, but also risks and vulnerability and, therefore, the reduction or fall in margins in operations, in a continuous process. Once a position has been achieved it will have to continue with the improvements because the environment forces with new and major changes.

The impact of technological changes affects directly the production, but also to the distribution of products and services, with new channels appearing that reduce or eliminate intermediation. It goes from a situation of strength, having investments in strategic locations, to be at one of disadvantage, with new competitors carrying out customers, products and services to new channels with zero investment and lower costs (Uber, Amazon and Airbnb are good examples).

The changes will be so fast that many companies will be expelled out of the market, so we will have to be ready to react to the new situation. This process will be much more aggressive for small and medium-sized enterprises, usually away from technological advances but it may be also an opportunity. Consequently, for the effect of having available tools to reach a wider market, and for the possible improvements in competitiveness resulting from digital transformation, provided that companies take advantage of their flexibility and adapt themselves to the new "virtual" approaches (companies with very small structure, concentrating their effort on their specialization) and considering:

1. To manage their activity with new technologies that allow disruptive transformation in processes, operations, products and resources and take advantage of cost reduction.
2. To simplify the "organization structure", with outsourcing services and focusing on core business to compete better and tackle more complex projects.
3. To ensure economic and financial balance.
4. To identify risks and to establish mechanisms for control and cybersecurity.

4 Management Information Systems

The Information System is key in this process. If the new information technologies are the support tool to carry out this process effectively, they need to be integrated into the Information system to achieve an action-oriented direction, facilitating a real time management system.

This will be the instrument with which Management ensures the creation of value, with continuous effectiveness improvements and operational efficiency, while carrying out economic and financial restructuring processes, reallocating resources, enhancing business lines, transforming business structures, or decreasing resources dedicated to others.

In the market economy, networked capitalism is also different, as politics and military action are. The Internet of Things (IoT), with billions of sensors and devices integrating software and intercommunication, allows us to improve efficiency in products and services in an extraordinary way, creating new business models, new supply chains, given an appropriate regulatory and legal structure.[4]

IoT devices not only need to be designed to include interfaces that people understand and know how to use but also with proven systems that allow

[4]The advances in mobility have been so spectacular that if Graham Bell came back to life he would be astounded by the telephone network and the mobile devices that we use today. But we are only at the beginning of a great change resulting from the existing technology and we can also think as Benjamin Franklin, which regretted to have been "born too soon" to enjoy the fruits of the scientific revolution. Google notifies that it is designing self-powered remote controlled vehicles that will be flown by and will be available within five years.

companies and their products, processes and services, to gather and process information in open communication, in an opportunity context.

Due to the networks, the nature of anything that is interconnected changes. A connected object, or an individual, a company or an organization gets a different personality. Relations give importance to any simple object, to any connected entity, but also make them dangerous what is the familiar and familiar what is dangerous and many of the problems that arise from being on the Net cannot be solved with traditional solutions or from comfort positions. Not everything is positive without forgetting that absolute access can also be destructive (Cooper Ramo 2016).

Our perplexity in an environment of connected crisis to understand why it happens and to do it before it breaks down our hardly homogeneous world, being aware that the only way to learn about networks is joining them.

Traditional business policy tools are unable to solve new problems, because they are on the Net, in the ecosystem, not only in the composition of factors. Old ideas, if implemented, would further worsen the situation and distort the solution to irreconcilable levels, not forgetting that inherent concept of hysteresis of cost also applies to technological fields. "It never goes back to a previous point, whatever the adjustment".

Companies throughout their experience accumulate rules, assumptions and beliefs of what the main reason for their success is and this "rucksack" conditions or even limits new initiatives. With the use of new technologies, these assumptions can be contradicted, and only a well-informed management A-team, knowing in depth those assumptions and mental acceptances, will try to obviate them and see what kind of effect a change can cause despite the conditioning of the inherited wisdom, experience and traditional knowledge. Few companies take the initiative to appoint a CTO and many fewer are willing to have external software engineering support to turn them into a cognitive company with increased intelligence.[5]

Cloud computing, big data, social media and mobility are the technological tools that will be the focus of the investments in the coming years, with digital transformation being the great facilitator of this change, but we have to overcome a very difficult barrier to change people's mindset.[6]

Circularity, energy and resource savings, waste disposal/reduction, virtualization-digitization and network optimization, connectivity and mobility are techniques to improve industrial competitiveness and create value in competitive environments. This improvement probably does not come from lower labor costs, whose savings are minimal taken globally, but from new disruptive designs in processes, components or products.

---

[5]Telephone companies, airlines and new logistic distribution companies are perhaps the ones that have made the biggest changes by breaking molds with previous situations knowing how to take advantage of a world of massive information, big-data, in a dynamic system.

[6]In fact, according to a recent study by Capgemini (2016), 33% of executives believe that digital transformation is a matter of survival, although only one in five Spanish companies has a suitable digital competence index, according to the Institute of Digital Economy (2016).

In order to achieve an efficient disruption, a global consideration is necessary to optimize, modify or speed up processes, both in geographic and virtual space, with disruptive opportunities in materials, processes and services, reconsidering the topological positioning and obviously taking advantage of a position/location asset.

It also requires identifying the elements of the processes, verifying which components are interchangeable and which are susceptible to incorporate software. All this action will facilitate modifications, substitutions or eliminations in the main industrial processes in a way we have never seen before, in order to optimize input and output, reduce costs, improve the acceleration of the production cycle and productivity.

The idea of interchangeable parts-elements, that was very effective in the second industrial revolution, made assembly lines and mass production possible. Nowadays, once a company has broken down its products and services into its key components and established interfaces to incorporate software, the next step focuses on how to use the software to assemble the components into new forms and products. This should be done taking care of security throughout the software in the system. All this process is based in the axiom that the whole is much more than the sum of its parts.

5 The Power of Network, Systems and Reflexivity

When we say that connection changes the nature of an object, it also happens for individuals, companies, agencies and institutions. The power in the network is formed by billions of points connected to a centralized core, in a large space installed between the center and the periphery. Power and influence will be in the near future, more centralized than in feudal times and more distributed than in the most open democracy. Distribution and concentration are the essence of power today and if the platforms were important, now the protocols are as important. The power of the network, with all the creative energy of a world full of devices and human power and violent divisions of old balances, can finally be a dangerous consequence of human selfishness and weakness, a debate that stands at the beginning of the revolution of the networks (Cooper Ramo 2016).

For Manuel Castells, "the society of networks represents a qualitative change in the human experience" that he calls "reevolootion" and considers that "we are witnesses of the birth of a new form of social movement" (Castells 2016).

Soros (2009) points out that there is a great deal of affinity between the systems theory and the reflexivity theory (the way in which human beings think determines the reality in which they live) that he proposes, since the reflexive phenomenon is complex.

The concept of reflexivity applies exclusively to situations that have participants thinking, making decisions and supported in two functions: that of understanding the environment in which they act (cognitive function) and that of changing the situation to reverse it in their favor (manipulation function).

In the real world, the opinions of the decision maker influence the course of events and the course of events also influences the opinions of decision-makers.

This circular and continuous influence turns into a dynamic feedback loop with distinction between objective (events) and subjective aspects of reality (thoughts). Even when there is only one external reality there are many different subjective points of view what define it, so that when we make decisions it is necessary to differentiate clearly facts from pronouncements.

However, even when knowledge exists and is available, its scarce application or misuse are noticeable. Society often proceeds to act by consensus, not by knowledge. This fact distorts and trivializes real problems of complexity, avoiding the difficult understanding of many phenomena, creating oversimplified models of the real world that exist only in the mind of the manager. Here lies the root of true incompetence and the true origin of manipulation. If artificial intelligence is integrated in the ecosystem, any subjectivity will be eliminated and the result of the decisions will certainly change.

Management theory is not expected to produce universal laws that can be reversibly used by management to explain and predict events. What is attainable in physics is not valid in social sciences, as they are reflective, so if the alchemists made the mistake in trying to change the nature of the base metals by sorcery and enchantment, they could have been successful if they had focused their attention on explaining the financial markets, as pointed out by Soros (2009).

For Beer (1975), man is a prisoner of his own way of thinking and his own instincts. His thinking machine, the brain, has been programmed to deal with a world that has already happened, but disappeared, and as we come from a journey in time for thousands of years accumulating experiences that has configured our genetic memory, which is our "rucksack." He also considers that while in the past the world was occupied managing things, raw materials and natural products, processed products, and using machines and tools, the today's world is characterized by the need to manage complexity; as a tool to achieve it arises the organization that deals with the present and its projection to the future whose change arise with explosive complexity.

Beer (1975) invokes science open to using deposits of knowledge, whether physical, biological or social systems, to define the model of organization and power: To measure and manipulate complexity, thanks to Mathematical Analysis, to design complex systems, using the General Theory of Systems, to design viable organizations with the support of Cybernetics, to analyze facts and data through Operational Research, to manage people with effective support in Behavioral Theory and to perform other actions based on Physics or Space Sciences.

It is the responsibility of those who direct institutions or companies to revitalize them avoiding their decay, renewing visions, missions and objectives, considering that:

- Time can play an important role of opportunity.
- The main instabilities of society can appear in an exoteric form such as "black swans".

- It is necessary to regulate ecosystems to induce stability by reducing risks where the dangers are evident.
- It is critical in the good direction of the companies how the information merchandise is handled, creating a meta-language to be interpreted uniformly by the ecosystems.

Institutions with new metrics and new organizational structures also need new strategy and new process of making decision, with fast dynamic answers, self-regulation and control. Possibly governments are the institutions that need more reforms under these cybernetic standards and it should be imperative that public channels of communication should not be allowed to distort available visions and by governments to implement effective cybersecurity systems that guarantee confidence in the use of networks as well as avoiding naive economic theories that help create unstable economies. While the government has become less powerful in the wake of colossal economic interest groups (large financial institutions, technology companies, energy companies, etc. operating in a globalized world), new tasks such as environmental protection, cybersecurity, space adventure, etc. will require more government, but with a different way of governing, and of course with a vision of international alliances. Decision makers need be connected with the reality ("wake up and smell the coffee"). A functioning society must be able to organize the tangible reality of the social order and no society can function unless it facilitates the individuals who form social status and functions to legitimize social power.

Crises are not "black swans", rare and unpredictable events, but "white swans", current events, although relatively infrequent, that even follow predictable patterns. It is even admitted that economic movements operate in cycles so that the delayed information is also incorrect and which represents some trend already overcome without being recognized. Some of the great impediments to achieving effectiveness are yesterday's issues, which condition and limit our vision and distort reality, seeing what has already been, not what is and will be.

6 Change, Complexity and a New Mindset

Change joins complexity and requires leaving behind what is not productive and focusing on doing what is really productive. The only way an institution can maintain continuity is by maintaining innovation in a systematic and organized way, creating a structure that acts as an organized destabilizer, managing change. To have stability and cohesion, a society needs dynamic imbalance. The weight lies in communication as well as effective knowledge management.

The more organized a change-leading institution is, the more it will need to establish internal and external balance coherence and an ongoing transparency effort to incorporate rapid change and achieve continuity.

Change cannot happen without the structure of an ethical support: humanity needs to formulate a new ethics relevant to the problems facing our species today and whose ethical basis is based on the truth and not spurious compassion of the real or "imagined" reality. In turn, the fear of failure has penetrated into society to the point where there can only be winners if there are losers. And it should not be so, so we need to recover values of fraternity along with recognition of merit and effort. Accepting this ethics assumes that:

- Technology recognizes humanity as a whole.
- That society needs to undertake responsible consumption.
- That society, in order to preserve nature and avoid threats to personal freedom must reduce or eliminate waste.
- Real-time control. "If things change too fast, management needs instant information." If information comes late, decisions are worse than irrelevant. Only thought can surpass the speed of light, but we need quality information and processes that respond with agility to our decisions.

Hence the need for adequate control, especially in an interconnected, open and therefore fragile vulnerable world ¿Could be a solution the nomination of a Web-Doorkeeper?

The manager with a new mindset has to think digital, even if he acts analogically. This means using technology to look at real data, follow interactions with different agents and extract information to better serve people. It requires clear and precise thinking rather than resting on rumors, habits and prejudices. Acting analog means putting a personal touch. No company acts without a high degree of analogy, regardless of their product, marketing or positioning on the web, and even when everything is web browsing, e-mails, sms, WhatsApp, etc. and to avoid the easy abuse of the digital, we must keep in mind that we keep on living in an analogic world. However, digital is not optional, digital transformation in business and society has come to stay. Digital natives are joined by digital immigrants with a vocation to be up-to-date and proactive,[7] knowing virtual does not exclude the analogic (Kawasaki 1999).

In society and today's organizations, people act more with knowledge than with skill and as the skill changes very slowly, knowledge becomes obsolete quickly, and it is crucial to reinvent itself periodically.

In a free society the public welfare must always rest on private virtue. In the company, each manager must know that the public benefit limits his own interest and with a broad vision and as a rule of conduct must be able to do what is good for the social welfare, and to make it real is the first duty and the legitimate basis of leadership.

As a rule, theory does not precede practice. Its role is to structure and codify the already proven practice by turning what is isolated and atypical (from the exception

[7]Proactivity means taking the initiative, being aware of the responsibility, and behaving according to our decisions, not our conditions.

to the rule with system) into something that can be learned and taught and especially generally applied. Considering that we live in an asymmetric society only using sophisticated means of control new catastrophes can be avoided without interrupting advance and progress.

About the Information System we also need to know how it behaves, so we could use some reference pattern that will serve as a model and see what happens after a small sudden change in the inputs or conditions of the environment (*Ceteris paribus*) and delimit their consequences as well as the delays in the decisions and actions and the policies that govern the operations that are carried out.

It is necessary to design simulation of management policies to give optimum control to the operations, the times of action and response, exploring the effect that the reduction of delays in compliance of operations and availability of data has in the operations and to see its results.

In order to configure a dynamic model for company simulation, it is necessary to describe adequately the real system it represents, and to obtain the data according to its basic characteristics (All components of a company's day-to-day management) along with other intangible elements such as the likelihood of reach results, development and innovation, the level of consumer response to advertising and market expectations, etc.

Many of these data, in principle unknown, can be estimated by establishing hypotheses, using methods that combine real data with intuitive factors and their incorporation identified in the system helps to reduce the uncertainty margin and sometimes with very effective results. But if it is determined that the system is very sensitive to the value of the factor, every effort must be made to determine the correct value. The risks of not knowing, acting with ignorance, increases daring and brings out the consequences. It is much better a good estimation (which has been thought with reflection) than an actual result achieved with delayed or biased or poorly drawn. Taking decisions based on accounting information, which requires documentary rigor, involves acting too late or not acting with opportunity. Information for management is different than accounting information, although required for legal supports.

As can be seen in the Information System, the role of the Control System can be very diverse, so it will have to be adapted to each business case. The new technologies also facilitate the creation of external platforms (outsourcing) that integrate their use and support a business process. As an example of application to the business-to-business environment projects are being developed with services in order to facilitate the Outsourcing of products and services that do not constitute the "core" or form part of its general services, to:

- achieve better profitability by reducing structure.
- provide high quality information for the contracting/management of the requested services.
- incorporate new information technologies in conditions of competitive advantage, in areas of mobility.

In any case, outsourcing needs to be performed with special security selecting the supplier incorporating the same levels of control and cybersecurity as the company establishes.

7 Conclusion

In order to ensure success and avoid digital risks we must take measures according to the complexity of the systems. Some of these measures are:

1. The first step starts with the strategic definition of the information system designing this in such a way that incorporates controls and firewalls in its core. Responsibility of management even when naming a CTO.
2. Incorporate into the Information System a risk management system that integrates digital risks.
3. Protect securely controlled data storage by encrypting essential information and using fragmentation mechanisms that make misappropriation almost impossible.
4. In the same way to establish controls in processes, input and output, designing safe mechanisms to prevent manipulations. Covering not only the company but also customers, suppliers and institutions connected, incorporating control of operations to verify its authenticity and those who operate with them.
5. Appoint a "Web Doorkeeper" that controls all networks and their interconnectivity. Network audits are essential, especially the software that constitutes the core of the information system and anyone who has access to, interfaces with, or can perform actions that compromise the company. The quality and trust of those who collaborate with the company is essential.
6. To seek from the telephone operator security services with prompt report of any anomaly. The opening of connectivity (and mobility) has to be limited to authorized applications, developing, whenever possible, customized applications, commissioned to professionals with proven experience.
7. When implementing IoT take special care that the software has the quality and protection that prevents "leaks" of security.
8. Be careful choosing social networks, which are more a communication problem of Corporate responsibility than a digital strategy one. Social structures are those which we are part of, be companies, universities, associations, clubs, etc., and are temporary collections of relationships that can change at any moment. Networks facilitate both concentration and distribution in social structures, with tension among them and their design, activity and control make up the real world.

9. In ecosystems (group of participants interconnected and open with a common purpose), who to trust and who we are connected to are the same.
10. Trust is what any company needs to act in a competitive environment where you cannot go back while going forward. You cannot go back from Google to the Encyclopedia Britannica.

Creating a climate of trust both internally and externally is the main function of company management. At the internal level, by drawing up a Code of Internal Regime that recognizes and commits all the people of the company. At the external level, establishing Corporate Responsibility documents that notify all interested parties of the company's position and the firm audit and control attitude towards any external attack.

Both are crucial. In the internal level it is the only way to ensure personal commitments and cover the new criminal law that places the company and its managers at the first level of responsibility for improper actions of its employees but demonstrates that it takes measures to avoid it. In the external area giving security of information to customers, suppliers and other related third parties is the guarantee.

In order to obtain reasonable security, it is advisable to rely on the proposed figure of a new "web doorkeeper" that controls networks, protects against unauthorized access or extractions of information and prevents feared "universe creators" from acting, opaque form but invisible, skipping rules and evading national and international legislation. We are facing a danger that has never existed before, to be virtual dominated.

But only one who has lived a global experience can have a conception of the new digital world and its interrelations in it, with billions of connected users, artificial intelligence computers, online commerce networks, and to understand what happens being connected to the network, being part of it, being part of any ecosystem.

Although we know that the network is controlled by human beings anywhere, we need to know who they are, and in some cases what values move them. Each person is today a node in an interconnected network, surrounded by devices and black boxes. But we do not know how or why they work or the values that make up the group that directs them, understanding it demands a new attitude in a digitized world, breaking with the established dare to know and not interrupting the advance at all. Therefore, we need to ensure our safety by incorporating a proper cybersecurity strategy.

References

Beer S (1975) Platform for change. Wiley, London
Castells M (2016) La sociedad en red: una visión global. Alianza
Cooper Ramo J (2016) The seventh sense. Little Brown, Boston
Gates B (1999) Los Negocios en la Era Digital. Plaza & Janés

Heck S, Rogers M (2016) Resource revolution. Amazon Publishing, Seattle
Kawasaki G (1999) Rules for revolutionaries. Harper Business
Soros G (2009) Lessons on theory of reflexivity. Seminar in Budapest
Welch J (2002) Hablando claro. Editorial Vergara

Author Biography

Manuel Gago-Areces is an economist by the Complutense University of Madrid, and has several other degrees, including a Technical Systems Superior (engineering systems) by Instituto de Informática—University of Madrid—and PADE-IESE, Navarra University, as well as a diploma in National Defence by the CESEDEN (Spanish Center on Higher Studies on Defence). Manuel has also attended courses on Planning and Systems Support at the California University, in Berkeley, and in Boston University. He was also a Lecturer at the Autonomous University of Madrid; he has been the General Manager of Tabacalera (currently ALTADIS), member of the board of the Pascual Milk Group, and Director of the Centre for Data Processing of the Banco Industrial de León. Among other honors, we may mention President of Board of Directors of AED—Madrid (Spanish Association of Managers) and at the present, Manuel is President of NEO SOLUCIONES INFORMÁTICAS, S.L and of MARKETEAM OPTIONS, S.L as well as Vice-president of CEDE (Spanish Confederation of CEOs and Directors).

Impact of Cyberspace on Individual Safety and Group Security—A Human Developmental Psychology Approach

Marzanna Farnicka

> *It has become appallingly obvious that our technology has exceeded our humanity.*
>
> Albert Einstein (2010)

Abstract This chapter focuses on the impact of cyberspace and cyber-activity on individual and small group security. The analyzed issues include problems such as losing social skills, cyber-violence and challenges or profits (multitasking, new way of thinking). The author describes cyberspace users according to developmental theories as a specific developmental context. Some psychological changes in the human mind caused by cyber-activity are presented, as well as their socio-psychological and cognitive-emotional consequences. Next, an analysis is carried out taking into consideration certain psychological conditions (attachment, temper, own experience) and motives for undertaking cyber-violence and cyber-bullying by young people. In the conclusion, programs for mental health protection and improvement for psychological safety for cyberspace users are presented. Two curricula aimed at increasing work effectiveness and lowering the costs borne by individuals who spend a lot of time in cyberspace are considered.

Keywords Cyber-bullying · Challenges of cyberspace · Brain · Social competences · Youth

---

Submitted: 8.9.16; Accepted: 3.10.16.

M. Farnicka (✉)
University of Zielona Góra, Zielona Góra, Poland
e-mail: m.farnicka@wpps.uz.zgora.pl

J.M. Ramírez and L.A. García-Segura (eds.), *Cyberspace*, Advanced Sciences and Technologies for Security Applications, DOI 10.1007/978-3-319-54975-0_6

1 How to Describe the Modern Developmental Context?

New technologies make it possible for us to observe daily the globalization processes[1] that are taking place. Availing ourselves of the resources of the 'global village' has become our everyday practice. The world has shrunk—socially, economically and culturally. We are able to do our shopping in various time zones on various continents and track the delivery 24 h a day in cyberspace, understood as the space that has been created and is still being created by new technologies (particularly the Internet). However, one should add here that the process of separating the traditionally understood time and space works both ways. Thanks to the use of cyberspace, today the world may not only 'compress'—coming closer, but also 'expand'—become more distant. Network participants can create their own worlds to suit their needs. The created worlds may be alternative to reality, and the network user may switch back and forth between the real and the virtual domain while retaining his or her core identity. Somebody may not feel psychologically attached to the actual physical location where that person is staying; on the other hand, someone staying physically close to us may actually live in a world that is unattainable and distant, because it is artificially created.

Another problem that has been manifesting itself is the greater and greater diversification of societies and distance between individuals measured by their competences and access to new technologies (Bauman 2011). The lack of readiness and competences for living in a globalized world may cause a globalization shock leading to apathy and resistance, a certain aversion (Nieman 2011) or even a loss of access to culture (Salzman 2001).

Fast technological changes drastically alter the conditions of human development. As the life environment has been changing rapidly, recognized models of development (e.g. the adequacy of the concept of developmental tasks: Arnett 2007; Baltes et al. 2006) and methodological approach (Mackenzie et al. 2005; Trempała and Cieciuch 2016) or notions used in psychology (Sampson 1989; Kowalik 2015) are being questioned. Developmental psychologists undertake the problem of insufficient knowledge and lack of efficient tools for the measurement of psychological processes connected with functioning in a world that is variable and unstable in different contexts (amusement, virtual, family life, professional life, etc., Farnicka and Liberska 2016). In this aspect, three questions have been asked:

1. What happens with the experiences of an individual transferred from one context into another and the eventually triggered or distorted intra- and interpersonal processes? Is it possible to close the experiences in one context, or is it, rather, that an individual transfers them and has to deal with them in a new one? From the developmental psychology perspective, transferred experiences that

[1]The term 'globalization' was introduced in the theory of culture by McLuhan (1960) to describe changes connected with the emergence of new media. Their use resulted in the subjective process of shrinking of the social space and increased velocity of social and economic interactions, which was named globalization or the 'global village' effect. Giddens (2008) pointed to the separation of time and space and uprooting of social institutions as indicators of that process.

have been worked through change[2] the structures of the mind and, hence, modify the perceptions, appraisal and evaluation of consequent situations.
2. What appears at the meeting point of these contexts, do they overlap, is an individual aware of crossing their borders or is it a rather smooth process? What emotions accompany this transition? Which competences are needed? What happens with the identity of an individual, including his or her sense of continuity in time and space (their place in the physical space, social place, interpersonal relationships and the world of emotions)?
3. What is and what should be the subject and the object of psychological research: observed behavior (according to games theory), genes, MRI potentials (according to neuropsychology), or the internal world of an individual's experiences (according to personality or developmental psychology)? The issue of the approach: should the studies be focused on the subject, the dynamic interaction, outcomes or on the whole system?

One of the ways to tackle and solve the problems discussed above is based on the contextual approach. It is worth reminding here that in the contextual approach *developmental environment* and *developmental context* are two different things. Developmental environment comprises all the material and non-material elements that an individual potentially has access to and by which he or she is potentially influenced. The moment when those objects, events or situations enter the psychological reality of the individual, they become the context of his or her individual development (Bronfenbrenner 2005; Brzezińska 2000; Farnicka 2011; Furmanek 2005; Kowalik 2015; Lerner et al. 2005).

Berry (2013) presented the model of analysis of three basic characteristics of individual functioning in a social world. Those are: attitudes of cognitive processes, social competences and goal-directed social activity. Another (older) proposal is the method of systemic description of complex psychological phenomena in the penta-basis approach. Ganzen (1984) proposed that the functioning of a given *psychological object* (*process*) should be described in relation to the following four dimensions: time, space, energy and information. With the use of that method one can describe interactions of each dimension with the object (process) and the emerging relations structure based on five elements. Approaching the process from the developmental perspective, those elements interact in time with the object and therefore researchers are able to observe and analyze the character of those interactions (at a given moment in time t, e.g. t1 or t2) or analyze the changes occurring in selected dimensions (e.g. between t1 and t2). Each *object* has its intrinsic characteristics, which determines its mode of existence in each of the dimensions.

The time dimension may describe the current status of the object and its developmental tendencies. One should remember that time is a multidimensional phenomenon. Various time dimensions are recognized for an individual human being or for a given culture: time may have a biological, historical, or social aspect or serve to differentiate

[2]According to the J. Piaget model of structuration and restructuration of psychological processes (1972).

between orientations towards the past, the future, or the present moment. However, time is not only understood as the dimension that describes one aspect of the developmental context (Trempała 2002). Time is a universal unit/measure that we devote to performing something. It is within a given unit of time that one can observe variety vs constancy and regularity versus randomness of occurrence of certain phenomena. The way of understanding and the attitude towards the phenomena observed within specific time may be characterized with the use of such dimensions as: flexibility, quick pace of performing tasks, undertaking risks, welcoming change versus conservatism, preservation of law and order, traditions, caution, repeatability.

The space dimension reflects the situation of the 'object' in relation to other objects. This dimension not only points out to physical aspects of space, such as material conditions, locations (of events or tasks) and their specific nature, but also to psychological aspects of space, such as closeness versus distance to others, openness to new experience or closing one's space, control or lack of control over space, loneliness or the presence of others.

The energy dimension indicates the nature of mutual relations between the 'object' and the environment and other dimensions. Various aspects of that dimension describe energy connected with: involvement versus apathy (succumbing to the course of events or even inertia), activity versus passivity, effort or resignation. The way of communicating with other objects is also characteristic for the energy dimension. One can observe that openness and permeability of boundaries may be used to describe space, but it may also relate to the energy dimension—if that openness or closing, permeability or its lack occurs as a result of actions (energy use) of an individual.

The informative aspect depicts and reflects cognitive orientations and the manner of analyzing the environment. That relation line reveals the manner of functioning in reality in the following dimensions: separating—consolidating, looking for similarities—looking for differences, analysis—synthesis, specification—generalization. Those strategies characterize both the ways of organizing and processing information coming from the outside world and from inside a given psychological system and the undertaken and analyzed activities.

The presented dimensions have already been used for the analysis of various psychological phenomena, such as creativity, needs, personality, mental health, or some complex issues such as the educational system (Rongińska 2013). Presented below is the systemic analysis of the functioning of Digital Immigrants and Digital Natives, as examples of the individual's strategies (and their changes) of actions undertaken in order to best adapt to the changing life environment.

2 Digital Natives and Digital Immigrants

Internet users are grouped into various categories according to the amount of time spent on the Internet or the manner in which they use it. One such criterion is being part of a certain generation: Millennials (people born on the turn of the 21st century and later), Generation X (people born between 1965 and 1980), or Generation Y (people born between 1981 and 2000). Those who were already born into the cyber world (i.e. the Millennials and Generation Y, people born after 1981) are Digital Natives; those who grew up in the world without cyberspace (i.e. born before 1981 —Generation X and older generations) are Digital Immigrants (Small and Vorgan 2011, p. 46). With the employment of this criterion, the very birth date (time) determines (cyber) space as the development environment and the manner of functioning in it (familiarity, being 'at home' versus distance, being a guest). Individuals from those two groups (worlds) use new technologies, but their usage has a different energy aspect (proactive vs. reactive approach), information dimension (looking for differences and new things vs looking for useful patterns and heuristics); the groups also differ in the sense of belonging to certain space and openness to its change. The systemic description of their functioning is presented in Fig. 1. This diagram was developed to depict the analysis of differences in cyberspace usage made by Small and Vogan (2011).

It is worth mentioning here that due to the time criterion, that division will soon become invalid, and new analysis criteria will be required for the classification of Internet users. Still, the opportunity to watch and analyze the functioning of two different generations according to the time of birth criterion is a fantastic opportunity offered to us by the creators of digital technology.

Energy 1. Involvement, pro-activity Readiness for change Low costs of using and experimenting	**Information** Looking for differences, novelty-seeking Looking for experiences, fun, satisfying many needs Life environment, analyzing details	**Energy** Reactivity, following Accepting, receiving High energy costs when changes occur	**Information** Levelling out differences Looking for usefulness Looking for models
Digital Native		**Digital Immigrant**	
Time Since birth Adequate speed – being up-to-date and on time – keeping up Risk; flexibility; novelty-oriented	**Space** Familiarized – at home. Open to new things, welcoming change Proximity Social and personal dimension Illusion of control	**Time** Limited – from a certain point in time Too fast or trying to catch up with changes Order, caution, fixed routines, rigidity	**Space** Guest: caution, avoiding change Distance Various dimensions – often things that are necessary New, surprised

Fig. 1 Analysis of two types of cyberspace users with the employment of the penta-basis method

3 The Consequences of Using Cyberspace for Individual Mind

Digital usage has not only altered the ways in which we communicate, organize information, search for knowledge and accumulate it, but also modified our time-spending habits. As a result, we witness changes not only in the way people think about the world (content), but also in the thinking process as such. What follows is the change in the way people feel and also the way in which they organize their functioning in society. Rapid, dynamic changes allow human development researchers to observe coping strategies adopted by two consecutive generations. Thanks to the creation of cyberspace, something technologically new, a generation gap has occurred in terms of ability to use that aspect of space. The emergence of that difference has made is possible to ask questions about the characteristics of the adaptation processes and adaptation skills that prove the most effective in face of new challenges for the generation of Digital Immigrants, and also to compare classical knowledge about various aspects of behavior with the changes observed in Digital Natives (Chuderski 2002).

Table 1 below presents differences observed by various researchers in the functioning of people using cyberspace as the main life environment and for those for whom that space is less natural. The material for the differential analysis was collected on the basis of the work by Small and Vorgan (2011). Due to the fact that the present-day science and norms were developed in the 20th century, the comparison of changes always relates to a certain 'earlier' time, adequate for the generation of Digital Immigrants. The presented differences depict changes in ways of functioning in space (material and social), using time, specific attitude (information) and energy involvement.

The analysis of the differences in human functioning that we may observe thanks to the existence of cyberspace is a source of questions concerning deeper consequences of that phenomenon. Analyzed below are selected changes that may be examined as results of transformations in the surrounding world caused by the technological leap.

4 Changes in Cognitive, Social and Physical Functioning

As an average the generation of Digital Natives spends about eight and a half hours in cyberspace,[3] of which 6 h are spent on the Internet; the remaining time is devoted to using the phone and playing games that do not require constant on-line presence. People who use modern equipment to communicate have been observed

[3]Data taken from Mobirank (2016) (https://mobirank.pl/2015/09/12/czas-spedzany-w-internecie-na-swiecie/). The study was conducted in 2015 in 34 countries worldwide.

Table 1 Analysis showing differences in functioning of individuals in the digital age and before it

Area	Before 1981	At present
Space-related habits	Reading books, magazines, watching TV, contact with nature, physical activity	Drop in those activities, looking for information on the Internet, contact with technology, decreased physical activity
Social functioning	Face-to-face meetings, personal involvement Building relationships—long-term processes, e.g. friendship	Speed; e-commerce, e-shopping, e-conferences, information on goods taken from the net without personal interaction Building cluster-type relations, task-oriented or ad hoc/occasional groups
Social contacts	Face to face, by phone	Networks, social networking sites, e-mails
Cyberspace control and energy	None	Addiction, compulsion
Attitude to technology—information	Concentration on utilitarian aspects Technical novelties feared, as they disturb the usual mode of functioning	Using gadgets, electronic toys New things welcome with interest
Functioning mode Values and gratification Learning	Linear performance of tasks Long-term gratification as a value Learning based on memory and concentration	Multi-tasking, parallel processing Short-term, instant gratification prevails Learning based on attractiveness and change

to develop stenographic skills and to understand and use a new language that is a lexicon composed of abbreviations and emoticons.

It has also been observed that computer games influence cognitive processes of their users. Peripheral vision and attention switching function are improved, multi-tasking dominates, mental reactions and decision-making processes are speeded up (Small and Vorgan 2011, p. 43; Chuderski 2002).

Another noticed change is a different mode of cognitive functioning in the learning process. Digital Natives often complain about unattractive form of traditional teaching. Single- or even dual-channel message mode employed by teachers (e.g. book or presentation) is boring and too slow for their expectations, which is connected with excitement seeking and concentration time (shorter attention span). A following example may serve as an illustration of young people's needs connected with content attractiveness: while watching an educational programme on TV, Digital Natives are also listening to music, handling their phones and taking part in an online chat, because as recipients they know that the motion picture can be stopped at any moment that is of interest to them. This style of learning provokes to the question concerning the processes involved in this activity: these are processes of identification, analysis, synthesis and thinking whether only the recognition (identification) processes[4]?

Those transformations, connected with new communication possibilities, have brought about changes in the frequency and form of face-to-face social contacts and the time devoted to cultivating bonds and building social competences.

A new style has been observed in interpersonal and intergenerational communication, based on texting, e-mails, sporadic necessary telephone calls to family and friends; also, certain anonymity, little activity related to social co-operation and overcoming frustration in relations with others from the so-called 'group of further friends'. The possibility of instant gratification of social needs on the Internet instead of regular social training in a given group of people in fixed time sequences (family rituals) is a threat to the development of communicative skills, emotional intelligence that is based on recognition of multidimensional non-verbal signals, as well as self-confidence in intimate relationships.

A dropping number of interactions between people, and particularly between parents and children, undermines the process of transmission of culture and prevents intergenerational support, which is detrimental for long-term functioning of relationships, families and groups. Being in a long-term relationship requires such traits as empathy and ability to cope with emotions, frustrations and mutual interdependence.

The time spent on offline physical activity undertaken for pleasure is also getting shorter. This phenomenon is illustrated by the latest experience with a game that forces physical activity of young players.[5] Young people spend hours outdoor solving mystery after mystery. However, the motivation behind that is not the pure pleasure derived from physical activity. A digression can be made here that the subjectivity of activity and will becomes an object in a game. Still, cyberspace is not the only place in which, by individual decision, subjectivity is treated as an object (Frankl 2006; Fromm 2005).

5 Functional Changes in the Nervous System

Multi-tasking, quick changes of stimuli and a high level of neural activity force our brains to adapt to such a functioning mode. In developing new connections and changing the frequency of activation of certain areas, we adapt and tune to the challenges faced by individuals living and working in cyberspace. For the brains of Digital Immigrants, those changes have the adaptive nature and are proof of our learning capacity and neuroplasticity. For most people, those changes take place

[4]The Problem of impact of the time pressure on the processes of thinking and decision-making was developed and analyzed by Pelucchi and Sillari (2016).

[5]Young players are looking for hidden objects or figurines with the use of maps. In order to win points, they need to be physically present in the place marked on the map (which is verified by GPS, One of games in this type is Pokemon Go).

imperceptibly. However, some people experience the so-called 'burnout', manifesting itself in attention deficits (also called distractedness).

Different processes occur in the generation of Digital Natives, for whom such a mode of functioning is a natural alternative that they practice since birth. In their case, researchers focus more on studying the development of those neural areas and connections that are rarely activated in cyberspace, and had earlier been developed in off line activities, mostly during social interactions and solving complicated tasks in memory. One of the problems that may serve as an example, is social spontaneity and the ability to empathize. Speaking from the neurological point of view, this relates to the simulation of the activity of the amygdaloid body responsible for initiating and maintaining eye contact. Research by Ybarry indicates also that everyday social contacts are also crucial for the development of memory, not only for the elderly but also for young people (Small and Vogan 2011, p. 179).

Brain activity studies with the use of magnetic resonance carried out with test participants online or using digital technology revealed the manner in which neural activity is activated and inhibited. Certain characteristics and slightly surprising results of that research are presented below.

According to research carried out up to this moment, the analysis and use of both spoken and written language activates the Broca's area in the frontal lobe. Meanwhile, writing and reading emoticons activates primarily the right inferior frontal gyrus previously associated with non-verbal communicative skills (Small and Vorgan 2011, p. 28).

Due to quick gratification of needs on the Internet and the possibility of providing stimuli that match expectations, the limbic system is activated. Frontal lobes and parietal cortex, i.e. brain parts that help us plan, control and inhibit cognitive and emotional arousal, are less often activated.

Questions about the consequences of presence of little children in cyberspace resulted in the emergence of hypotheses connecting a different mode of their cognitive, social and emotional functioning with changes in brain functioning. Problems so far labelled as concentration deficits or psychomotor over excitability disorders may be treated as a new model of cognitive functioning.

An example of such an understanding of the changes in the way in which children function is the phenomenon of 'Indigo Children' introduced by Carroll and Tober (1999). The term applies to children who are highly talented but socially maladjusted. What makes those children unique, apart from their problems with discipline, observing norms and respecting authority, is their exceptional creativity, intelligence, intuition, exaggerated self-esteem, and frustration and boredom connected with functioning in traditional social systems (school). The numbers of such children are growing and this was probably the reason behind linking their exceptional features with functional changes in the brain resulting from the use of modern technology. Classically understood, the symptoms listed above would suggest cognitive deficits, hyperkinesis and behavior disorders, as well as disturbances from the autistic (Asperger syndrome) or psychotic (hebephrenia) spectrum (ICD-10–CM

2016). A frequent danger resulting from the search for new functioning models is the delay of the correct diagnosis, as well as therapy that would enable a child to function with success in the social world (Small and Vorgan 2011, p. 105).

6 Influence of Individuals on Cyberspace

The emergence of cyberspace has brought about the occurrence of such antisocial phenomena as cyber-bullying, cyber-violence and cybercrime. Mapping a new developmental environment reveals the existence of numerous threats connected with those phenomena. As cyberspace is programmed—settled by people or human-made programs, this section presents individual features and motives of people who engage in activities that may be classified as cyber-bullying and cyber-violence (Menesini et al. 2016).

Violence with the use of electronic media has become a social issue in many countries. A metaanalysis carried out by Gradinger (2014) shows that the motives of perpetrators of cyber-violence and cyber-bullying are most often anger, authority (sense of power), as well as free-time entertainment. Hitherto conducted research indicates that in case of combined motives (e.g. authority, power and anger) the risk of all types of aggressive behavior in all the traditionally recognized channels (verbal, physical and indirect) increases. When motives such as anger prevail, one can suspect that a perpetrator will choose a single channel considered most effective under the circumstances (Dixon 2011).

One may quote here the assumptions connected with the concept of frustration—aggression, i.e. that such behavior stems from reactive motives such as the attempt to cope with anger, or instrumental motives serving the purpose of achieving authority, power or affiliation (according to the assumptions of the concept of social learning to achieve goals). If the motive of play is involved, the development of such behavior is hard to determine, as very few studies have been conducted from that angle. So far both the need of play and boredom have been treated as specific motives for violence activity on the net, as in real life such motives are behind the inflicting of harm on others, and are treated as predictors of disturbances of psychopathic nature (Cheung-Blunden et al. 2008).

The research of Gradinger et al. (2015) indicates that the rise of bullying and cyber-bullying is parallel. This may mean that the reasons for both types of behavior should be sought in similar psychical processes of perpetrators and also in processes taking place in groups, or one may assume—after Mensini and Nocentini (2009)—that cyber-bullying is about the group's aggression towards an individual, but in a modified environment. Reports on cyber-violence show the consequences of such experience for psychical well-being, mental health, educational and social achievements, and interpersonal relations (Smith 2007). Negative consequences of behavior recognized as violating the norms and principles and encroaching personal boundaries affect not only individuals who are directly involved in the process, but also those who witness it.

The Czech researchers Machackova et al. (2014) draw our attention to the consequences of witnessing cyber-violence. Their study was conducted on a group of 257 school students aged between 11 and 19 (N = 257, age 11–19, average age 16, SD 1.9, 77% girls). The results reveal that in cyber-violence witnesses, the level of support for the victim was dropping along with the increased number of attacks and the duration of the process of bullying. The researchers called this phenomenon the specific 'spectator effect'. Protective factors in that situation were empathy, sense of self-confidence and good relations with the victim. Those three factors increased the probability of help being offered to the victim. However, as pointed out by Cheung et al. (2014), empathy is a specific double-edged tool that not always guarantees offering help to the victim, as due to their sensitivity level, emphatic individuals may have problems with undertaking actions aimed at getting the perpetrator into trouble. The research of Machackova et al. (2014) confirmed that proposition, indicating that the high level of the observer's empathy and the long-term process of violence cause lowering of the perceived attractiveness of the victim and, consequently, leads the observer to dissociate him- or herself from the victim.

Research by Farnicka and Grzegorzewska (2016), aimed at separation of individual traits responsible for the undertaking of aggressive behavior on the Internet among teenagers, has emphasized the significance of attachment type (mother alienation) for the undertaken risk behavior, both for the perpetrator-type and victim-type. The differences relate to the type of relationship with parents (secure or unsecure pattern), their own experience of being in the victim or aggressor role, and the level of hostility. A separate model of determining factors for indirect type aggressive behavior was also demonstrated (N = 120, age 16–19).

In the light of the presented results, the proper diagnostic process seems important in order to differentiate between assumed roles (aggressor/victim) and then be able to apply suitable measures. In the case of indirect-type aggressors, the focus should be on coping with hostility, self-control and emotional control skills and working on alternative ways of functioning in the family, particularly in the relationship with the mother ($R^2 = 0.084$, $F = 5.25$, $p < 0.001$, beta trust in the relationship with mother (beta = −0.209) and mother alienation (beta = 0.3)). It should be mentioned that, in the light of those results, a person using indirect aggression may not display aggressiveness in any other form (in line with the assumptions made by Gradinger 2014). This may mean that undertaking aggressive behavior on the net may have a stronger connection with the sense of rejection or unsafe attachment models. It may also point to the use of the 'cold' strategy (e.g. revenge), or acting for fun or out of boredom.

As for cyber-violence victims, research results find them most often to be victims of other types of violence as well. Therefore, prevention and therapy should take into account various reasons and motives of such behavior and should concentrate on coping with hostility, self-control and emotional control skills; such individuals should work on the ways to function in close interpersonal relations (in the family, particularly in the mother-son relationship).

The results of that research emphasize the significance of the relation and attachment bond for undertaking aggressive behavior aimed at harming somebody (even on the Internet). This stresses the role of socialization and the significance of interpersonal relations.

7 The Challenges—How to Protect Individual Safety and Group Security in Cyberspace

In everyday life, safety and peace are treated as one and the same thing, and considered to be taken care of by the state. On the other hand, individual safety—also important for people ensuring external safety—is one's individual concern. In a situation when the margins of safety (individual, group, physical, economic, international) are infringed, the existing homeostasis (balance) is shaken, resulting in what may be called a conflict, and an action necessarily follows which may assume the form of confrontation, mediation, negotiation, apparent activity or any other adequate or inadequate activity (Hobfoll 2006).

8 Changes in the Personal Safety in Cyberspace

Functioning in cyberspace is connected with many threats to safety and security of both individuals and groups.

The basic dangers from the point of view of an individual are total anonymity and loneliness on the one hand (as one can use his or her nickname or alias and be online at any time) and fear on the other hand. That fear is the fear of losing privacy and leak of data gathered on home computers, in the cloud or on social networking sites, or even fear of being rejected in the real or the virtual world.

Another threat is the power of influence of various contents that can be found on the net. Every now and then, reports occur in the media about strange Internet-spread diseases, hysteria cases or even Internet group suicides. One example of such mass hysteria may be the fear of catching the 'strawberries with sugar' virus. "Morangos com Açúcar" ('Strawberries with Sugar') was a popular teen soap opera broadcast in the early 2000s on Portuguese TV. In one of the episodes shown in 2004, the characters were infected with a virus that spread at the school they attended. The symptoms of the disease included skin rash, troubles with breathing and dizziness. After a few days the 'disease' was diagnosed in over 300 real (!) students from 14 different Portuguese schools. Some of the schools were temporarily closed and quarantined before it turned out that the disease was only virtual (Taler 2014). Another example is the setting up of online groups supporting suicide or self-mutilation practices. Young people share tips on how this should be done or even offer encouragement during online transmission. In 2013, Stephen from the

USA was broadcasting his suicide attempt for 40 min. During that time he was also staying in touch with other Facebook users (Daily Mail 2013). As Internet is used to exchange information and form suicide pacts, the Internet Safety Commission censors blogs, making sure no words such as ‘suicide’ or ‘death’ appear in the titles of blogs. For instance in South Korea search engines are programmed in such a way so as to prioritize addresses of counselling centers in search results when the word „suicide’ is keyed in (Bendyk 2005; Farnicka and Magda 2014).

As an Internet user has control over the content found online and the intensity of that content, we can say that such a person creates his or her world of values and accepted behavior and may fail to notice other aspects of reality or his/her own functioning. Such (self-imposed) limitation of access to information is a threat to building one’s identity, self-esteem and proper judgement of reality. An example of that problem is the popularity of various blogs and sites related to solving various life problems, from treating influenza to treating cancer and from passing a chemistry test to healing a broken heart. The links provided below show up as answers to the Internet search of ‘How to heal a broken heart’: 51 addresses on 5 screens have been found. The offered tips describe 3, 10, or even 33 steps to cope with grief following relationship breakup within 7–30 days (some of examples: www.lifehack.org/articles/communication/how-heal-broken-heart-and-the-science-behind-2.html, www.wikihow.com/Heal-a-Broken-Heart, www.huffingtonpost.com/joyce-marter-/broken-heart_b_4645774.html).

Internet blogs or forums are under no professional verification, being only controlled by moderators or administrators with regard to compliance with the rules. They are often hidden, not accessible for everyone (e.g. for parents or friends). At this point it is worth noting that the problem relates not only to the question of safety and reliability of information but the development of deeper structures of human psyche, such as awareness building and self-limitation of one’s development (narrowing the field of exploration by excluding other dimensions of life or functioning).

Excessive presence in cyberspace has got new negative consequences and poses threats to the safe functioning of individuals. More and more often doctors link sleeplessness and obesity to night-time Internet activity and the new pattern of day and night, activity and regeneration (Akmed 2010; Cogito 2011).

According to statistics, about 20% of Internet users meet the clinical criteria of addiction. In simplified terms, one can say that the addiction process relies on dopamine release in the situation of excitement and gratification (Jakubik 2002). Due to the fact that gratification on the Internet is irregular, strong gratification balances the influence of numerous disappointments caused by, for instance, losing a level in a game or receiving an e-mail with some negative information. As is the case with other types of addiction, Internet addiction involves mood swings, lowered tolerance to abstinence, withdrawal symptoms and relapses. The American Medical Association is currently in the course of studies to determine whether it is legitimate to officially recognize *Internet and computer games addiction* as a separate diagnosable disorder (Cash et al. 2012).

Another threat to individual safety of cyberspace users and also to the safety of small groups, such as families, is the threat connected with e-shopping, e-gambling and e-politics addiction. Problems with taking care of one's own money, control of spending or control of one's economic or political actions stem from the fact that both the value of money and the perception of the responsibility for one's words in cyberspace seems lower than that in real 'face-to-face' encounters (Burgess 2005; Falahati 2011; Tyszka 2005).

9 Improving Individual Cyber-Safety in Cyberspace

In the face of numerous advantages that cyberspace offers and also threats that it poses, some challenges are worth mentioning that may increase the level of individual safety. Users themselves are responsible for their own safety; also, people who organize time for others (parents, teachers, doctors, corporation managers, etc.) are responsible for ensuring the safety of those under their supervision.

The awareness of the impact of that developmental context on behavior should encourage steps aimed at enjoying its advantages along with the prevention of its negative effects. First of all, this means the recognition of the multi-dimensional aspect of human existence and the necessity to function in numerous relations in many worlds, both digital and traditional. Therefore, individual knowledge and attitudes towards new technology become important. Presented below are some aspects in which online safety levels may be increased to tackle such issues as addiction, anxiety, loneliness, communication and environment organization.

Individual competence should comprise fluency in digital technology usage (particularly in the field of data and content safety) and knowledge of threats such as that of developing an *addiction*. At that moment, an individual should recognize and be alert to addiction symptoms, such as preoccupation and over-involvement, thoughts revolving around online activity, lower tolerance to being offline, longer periods of staying online and looking for pretexts to get online, lack of control and negative feelings when network activities need to be interrupted, situations when any attempts to change the activity model (skipping online activity) lead to unrest, irritation, mood changes, sense of emptiness and in consequence the online presence is even longer (relapses—the yo-yo effect). The symptoms listed above decrease individual well-being and may be subjective. However, functional disturbances connected with online presence—such as a real threat of losing one's job, position, or important personal relationship, lack of promotion at school, hiding the actual time spend on the net, using the Internet in order to alleviate negative feeling or running away from real-life problems—should be alarming both for the user himself and for his family and friends.

Another effective method, apart from awareness, is the introduction of protective habits in Internet usage, including time limitation (protects against attention deficits and distractedness) and diversification of activities by including physical activity (protection against physical problems such as degenerative conditions of the spine, spine ache, sore eyes and obesity).

Loneliness, anxiety. Research indicates that the development of the sense of coherence and mindful approach to self-development processes are protective factors against disturbances that result from imbalance (Hobfoll 2006). Positive bonds and successes connected with one's own activity also have a preventive function. Social interactions are an example of such gratifying activity. The process of communication between generations is a natural resource. Conscious focus is required in building positive experience and relations between those two worlds and modes of functioning, through the support of natural ways of communicating or, if necessary, specialist training such as social skills training, assertiveness training, empathy and listening training, etc.

Organizing one's environment—early prevention, primary and secondary. At present, Digital Immigrants are mostly responsible for activities of that type. However, that responsibility will soon be passed on to the first generations of Digital Natives. Individuals from that group should be required to 'keep their eyes on the ball' and ensure a proper dose of social training and face-to-face relations for both children and adults. Shaping *social skills and personal competences* should be a mandatory element in the educational process, and should be verified (for positions connected with group work) and systematically developed in adult life. Without proper shaping of communicative skills and their development we deprive ourselves of tools for solving intra- and interpersonal problems. Also, a proper amount of offline training not only enables intergenerational communication but widens the developmental context, and thus helps us build a stable identity and self-image.

10 Changes in the Group Security in Cyberspace

Cyber-violence and cyber-aggression are often manifested as infringement of personal interests, dampening somebody's spirits or lowering his or her position in the group, all that achieved by using communicative techniques on the Internet. Most often such activities involve extortion, imposing something on the victim or ignoring him, depriving the victim of information, ignoring, belittling or exaggerating certain features, behavior or work results in a given social context. In this aspect, relevant research indicates that protective factors are personal traits such as realistic self-evaluation and the feeling of peace and happiness (Steffen et al. 2014). There is a slogan in English literature promoting self-improvement: 'stability of mind in unstable times'.

The synthetic activities suggested below may be viewed as measures intended to shape up protective factors by equipping cyberspace users with social competencies, particularly in the sphere of interpersonal communication, constructive conflict solving, assertive protection of one's rights (as a communication technique) and coping with stress, implementation of suitable norms and procedures that will actually prevent cybercrime. Presented below are the descriptions of two different solutions that may serve as the starting point for the development of one's own programs of increasing cyber-safety from the perspective of primary and secondary prevention.

11 The 'Keep It Tame' Campaign

Over the 15-year history of social media, reports about cyber-violence have been pointing to the consequences of that experience for psychological well-being, mental health, educational and social achievements and interpersonal relations. In 2012, a survey was made in Australia about the impact of social media based on a big media campaign called *Keep it Tame*. The Young and Well Co-operative Research Centre project's objectives were to change youth online attitudes and behaviors, so they are more respectful of others. The campaign, *Keep It Tame*, guides teenagers through a series of mock social media posts that will be run on websites they frequent. As things turn nasty, an animated creature slowly becomes more grotesque, highlighting the hurtful effects of the exchanges and ultimately encouraging people to act with respect. The programme is based on four pillars: right to respect, right to self-acceptance, right to seek help and aspiration to be integrated.

In the first step, project participants are taught how to promote respect on the net. They may take care of themselves by using the icons that signal 'nasty' information or by deleting people who damage their reputation from the contact list. In the second step, participants have at their disposal icons that cheer up and contain kind messages.

The leader of the researchers' team was Barbara Spears. She presented the first results in 2014 (Spears and Taddeo 2014). The research was conducted among teenage users of the platform (N = 345, ages 12–18). Results of the study indicated that adolescents who consider themselves victims of cyber-bullying demonstrated higher levels of fear and depression than their peers and evaluated themselves lower in social relations. Researchers observed the frequency of their use in Internet contacts. Studies carried out after one year and two years after the programme implementation, indicated that the number of cases of cyber-bullying and ordinary bullying among participants had dropped, fear and depression decreased, while self-evaluation improved.

What is worth noting is that the individuals with the best mental health, self-evaluation and self-confidence felt that the cyber-bullying problem did not concern them at all.

12 The GePePeS Project

The general aim of the presented tips to prepare curricula is to raise the competencies and operating efficiency of individuals engaged in cyber-security consolidation and safety-ensuring missions. The presented stages of programme cover two fundamental training areas which are indispensable for such people to perform their tasks aimed at stopping activity in cyberspace by Offenders and keep their mental health[6] (Farnicka et al. 2015).

Description of the proposal Conflictive, aggressive and violent experiences in all their forms are triggered and influenced by a multitude of internal (individual) as well as external (social) factors. Cognitive processes related to human perception, attitudes and socio-cultural norms are involved in this behavior. The people engaged in security oriented missions, through negotiation, confrontation or establishing and keeping order, must be able to recognize the stages of connected threats, and should be able to activate social activity and engagement of the environment around the actions that they undertake. Another important aspect of raising the competencies of staff engaged in building and maintaining peace is the ability to control emotions and to counteract pathological reactions connected with working under prolonged stress caused, for instance, by changing one's place of stay, encounters with strangers and hostile people, and enormous pressure and responsibility.

Aims The project focused specially on: 1. recognition, 2. understanding and 3. awareness training in building professional skills. The adopted model of analysis of documents and current methods from the perspective of three [time] planes will enable the comparison of causes and the determination of a strategy of conflict resolution despite contextual differences (culture, values, social expectations of a given group), and will enable the preparation of a new training path.

Programme approach The training curricula should be based on the same assumptions as the analysis of current training programs and conflict understanding and prevention, i.e. the systemic perspective and the interactive-dynamic paradigm (Lee et al. 2004; Meyer et al. 1993; Pervin 1993). The objectives of the programme should include the development of competencies and the improvement of efficiency through the preservation of resources of the individual in the workplace (Hobfoll 2006). A general assumption of Hobfoll's theory is that every individual has some resources at his/her disposal which he/she greatly appreciates and is inclined to protect and never lose. Stress is predicted to occur as a result of circumstances that represent: (1) a threat of resource loss, or (2) an actual loss of resources required to sustain the individual, and (3) the lack of any reasonable gain following resource investments (Dudek et al. 2007). Being under stress in the workplace can be

[6]GePePes—Curriculum for Resolution and Peace Keeping Personnel was prepared by researchers connected with the universities: Zielona Góra and Nebrija in 2015 as a Grant to H2020

understood as an effect of some environmental stressors, perceived by an individual as exceeding her/his abilities to face them (McAuliffe et al. 2007; Ray et al. 2008).

The aim of the project should be to maintain a good functioning despite the occurrence of stressful and frustrating experiences connected with the professional activity. According to the assumptions of Prochaska et al. (2008), who put forward their transtheoretical model of change, the impact that we call 'the programme' shall contain the following elements:

(a) widening the awareness—providing information on conflicts and their dynamics and the systemic approach to factors influencing conflict occurrence, providing information on stress disorders and manners of coping with them, as well as information on group communication and functioning within the group; and
(b) activating the processes of social support, a network of mutually supportive relationships, processes of involvement in one's work and control of own emotions, elements of monitoring and controlling the environment in which the working task is carried out. Also, the second module will include working with emotions and their control, and the issue of fixed thinking patterns connected with self-evaluation in the form of feelings and thoughts about oneself in connection with work.

13 Conclusion

The answer to the question of safety in cyberspace should begin with the proposition that cyberspace as such is neither good nor bad. It is an invention that offers humans the opportunity to explore the world in a new way. It is a gift, as was the print, radio, television, airplane or telephone. It inspires both fear and desire. The existence of that dimension makes it possible for us to learn more about ourselves and others and optimize our activity or enhance possibilities connected with science and gaining knowledge. In the history of mankind, particularly the history of science, man often faced new and potentially dangerous phenomena that later became familiar and proved useful, while still retaining their danger potential. By way of example one can mention the nuclear fission of uranium. The energy obtained from that physical reaction may be used for making nuclear weapons, but also for generating clean energy in nuclear power plants.

Digital technologies contained in cyberspace save time, increase productivity, open new ways in education and simplify many aspects of life (Garito 2008). When evaluating that life dimension, one should follow the principle of ensuring maximum possible safety rather than negating its significance. With such an approach, even basic threats of cyberspace become potentially bi-valued. Stoll (2000) indicated that the fundamental threats of cyberspace may be anonymity and accessibility. The results would manifest themselves as devaluation of close relations and closeness, sense of loneliness, atomization and lack of support. However, studies indicate that for some people (for instance for addicted people) this is indeed the case. However, it has also been noted that thanks to anonymity and availability of cyberspace many individuals find the courage to ask for help, openly admit to mistakes and receive support from various Internet forums (Huk 2007; Wallace 2001).

Another threat connected with cyberspace is that the anonymity and openness of the net poses a risk to data safety and offers unlimited opportunities for various frauds and hackers who give false information or use illegally obtained data and remain unpunished. On the one hand, everybody has got limitless possibilities of self-creation and opportunities to meet people from various cultures. This offers a chance to neutralize stereotypes. On the other hand, cyberspace involves a risk of falling a victim to various forms of abuse (pornography, aggression, theft, or false relationships) and the disseminated information or programs may pose a threat to basic security, even from the global point of view (García-Segura and Ramírez 2016). As early as 1999 Dyson pointed out that the Internet may become another area of the arms race, as through its reach it could become a powerful tool in the hands of anyone willing to use that channel (type of weapon) skillfully. Wallace (2001) provided a similar description of the dangers connected with cyberspace, stating that '*The Internet may be a machine of conspiracy, rebellion, resistance, as well as a propaganda of official authority or product'*, p. 307).

At that point, a question arises about safety monitoring on the net and fighting various pathologies, from cyberattacks on government and military data to money swindling and pornography. The main concern is how to stay ahead of this genuine race without compromising the value of privacy.

In 1994, Hughes pointed out that great technological inventions may alter the way in which societies function. Today, twenty years after the creation of conditions for living in a 'global village', one can say that technology is the outcome of development but also a starting point for new forms of communication, interpersonal relations, education, and shaping up of human expectations and wishes. One may venture a statement that cyberspace imposes a certain way of functioning (i.e. it is a manifestation of technological determinism) but on the other hand it is created by us humans and human-developed programs (sociological constructivism).

In this world (cyberspace) that was once new and unknown we now have certain well-proven patterns of behavior that give us the illusory sense of control and influence.[7] Specific heuristics were developed to minimize danger; some of them

[7]The issue of influence on and control of cyberspace safety and security has been widely discussed. The conclusion of these discussions is that due to speed, large number of users and potential inventors, people or staff who are responsible for cyber-safety depend on reactions of others who can spot the danger. The situation of a cyber-safety guard is similar to that of a ship captain—not only must they perform their work with commitment and in compliance with currently available knowledge and technology, but in order to avoid unexpected dangers they should pay attention to many co-occurring phenomena and be open to information provided by other specialists and from other perspectives.

being critical thinking, responsibility, caution, good user skills and knowledge of security procedures.

From the point of view adopted for the needs of this paper, i.e. the one of developmental psychology, in order to maintain a basic level of individual and small group safety on the net in the face of constant technological progress and changing life context today and in the near future, the development and preservation of autonomous mature personality ('reflexive construction of the self' according to Giddens 2001) and a high level of social and communicative skills (through education in the interpersonal context and digital skills) is necessary. Those two aspects of functioning are capable of keeping us at balance between efficient use and development of technologies and staying in touch with other people and oneself in the 'self-self' relationship. Underlining the importance of those developmental aspects (identity and maturity, as well as communication) is particularly important, as very soon the generation of Digital Natives will take over the responsibility for creating the conditions for their children's development and for transmitting knowledge and skills. For people raised in a different environment, direct interpersonal contacts may become a choice or option, not a necessity. This realization should inspire care for proper procedures and skills ensuring development in diversified life environments.

References

Arnett JJ (2007) The long and leisurely route: coming of age in Europe today. Curr Hist 130–136, March. Taken from: http://www.jeffreyarnett.com/articles.htm

Baltes PB, Lindenberger U, Staudinger U (2006) Life span theory in developmental psychology. In: Damon W, Lerner RM (eds) Handbook of child psychology, vol 1, 6th edn. Theoretical models of human development. John Wiley & Sons Inc, Hoboken, NJ, US, pp 569–664

Baumann S (2011) Collateral damage: social inequalities in a global age. Polity Press, Cambridge

Berry JW (2013) Achieving a global psychology. Can Psychol 54(1):55–61. doi:10.1037/a0031246

Bronfenbrener U (2005) Making human beings human bioecological perspectives on human development. Sage Publications, Thousand Oaks

Brzezińska A (2000) Społeczna psychologia rozwoju. Wydawnictwo Scholar, Warszawa

Burgess SM (2005) Money attitudes and innovative consumer behavior: hedge funds in South Africa. Adv Consum Res 32:106–126

Cash H, Rae CD, Steel AH, Winkler A (2012) Internet addiction: a brief summary of research and practice. Curr Psychiatry Rev 8(4):292–298. doi:10.2174/157340012803520513

Carroll L, Tober J (1999) The Indigo children: the new kids have arrived. Hay House, Carlsbad

Cheung-Blunden V et al. (2008) Paving the road to war with group membership appraisal antecedents and anger. Aggressive Behav 34:75–189

Cheung-Blunden V et al. (2014) Empathy: a double edged sword. Kultura i Edukacja culture and education 5(105):193–210

Dixon R (2011) Rethinking school bullying. Cambridge University Press, New York

Dudek B, Koniarek J, Szymczak W (2007) Work—related stress and the conservation of resources theory by Stevan Hobfoll, Medycyna Pracy 58(4):317. http://medpr.imp.lodz.pl

Dyson E (1999) Wersja 2.0: Przepis na życie. Prószynski i S-ka, Warszawa

Einstein A (2010) The Ultimate Quotable Einstein, Edited by Alice Calaprice, Princeton University Press, Princeton, New Jersey; New York Times, 1946 May 25, Atomic Education Urged by Einstein, Page 13, Column 6, New York. http://quoteinvestigator.com/2012/10/25/tech-exceeded/#return-note-4654-3. Accessed 29 Sept 2016

Falahati L (2011) A comparative study in money attitude among university students: a gendered view. J Am Sci 7(6):1144–1148

Farnicka M (2011) Przemiany realizacji zadań rozwojowych. Ewolucja czy rewolucja?. UZ Press, Zielona Góra

Farnicka M, Grzegorzewska I (2016) Intrapersonal correlates of aggression in adolescents: determinants of undertaking the role of the perpetrator and the victim. Curr Issues Pers Psychol 3:25–35

Farnicka M, Liberska H (2016) Stages and paths of aggression development—Knowledge that awaits being uncovered. In: Liberska H, Farnicka M (eds) Aggression as a challenge. Peter Lang GmbH, Frankfurt am Main–Bern–Bruxelles–New York–Oxford–Warszawa–Wien, pp 15–31

Farnicka M, Magda E (2014) (eds) Samobójstwo. Jeden problem, trzy spojrzenia (The Suicide. One problem—three approaches: psychological, sociological and resocialization). Oficyna Wydawnicza Uniwersytetu Zielonogórskiego, Zielona Góra

Farnicka M, Ramírez M, Cayón J, Fernández JC, Calleja G, Payá C (2015) Improvement of mental health and cooperation skills: a curriculum for peace keeping personnel. In: García-Segura LA, Martin Ramírez J (eds) Mapping the cyberspace. An emerging priority challenge. Universidad Antonio de Nebrija, Madrid, pp 36–37

Frankl V (2006) Man's search for meaning. Beacon Press, Boston

Fromm E (2005) On being human. Continuum, New York, London

Furmanek M (2005) Media i multimedia jako środowisko edukacyjno-wychowawcze. In: Izdebska J, Sosnowski T (eds) Komputer w życiu dziecka i obraz jego dzieciństwa (trans: Humana). Białystok, pp 17–24

Hobfoll SE (2006) Been down so long, it looks like up. Cognitive recalibration of group in response to sustained conflict. Palestine-Isr J 10:18–23

Jakubik A (2002) Zespół uzależnienia od Internetu. Studia Psychologica 3:133–142

Ganzen VA (1984) Sistemnye opisanija v psihologii [System descriptions in psychology]. Leningrad University Press, Leningrad

García-Segura L, Ramírez JM (eds) (2016) Mapping the cyberspace. Universidad Antonio de Nebrija, An Emerging Priority Challenge. Madrid

Garito MA (2008) Universities in dialogue in a world without distance. In: Guske I, Swaffield BC (eds) Education landscapes in the 21st century: cross-cultural challenges and multi-disciplinary perspectives. Cambridge Scholars Publishing, pp 355–368

Giddens A (2001) Nowoczesność i tożsamość. "Ja" i społeczeństwo w epoce późnej nowoczesności (Modernity and self-identity. Self and society in the late modern age). PWN, Warszawa

Giddens A (2008) Konsekwencje nowoczesności (The Consequences of Modernity). Wydawnicto UJ, Kraków

Gradinger P, Yanagida T, Strohmeier D, Spiel C (2015) Prevention of cyber—bullying and cyber-victimization: evaluation of the ViSC social competence program. J Sch Violence 14:87–110

Gradinger P (2014) Risk Factors for and Prevention of Cyber-bullying, lecture at XIX workshop social and media dimensions of aggression. Berlin, November 20–21

Hughes TP (1994) Technological momentum. In: Smith MR, Marx L (eds) Does technology drive history? The MIT Press, Cambridge

Huk T (2007) Edukacyjna wartość blogów. Chowanna, 63,142–152. http://bazhum.muzhp.pl/media//files/Chowanna/Chowanna-r2007-t2/Chowanna-r2007-t2-s142-157/Chowanna-r2007-t2-s142-157.pdf

Kowalik S (2015) Globalizacja jako kontekst funkcjonowania psychologicznego ludzi. Nauka 1:7–37

Lee T, Mitchell T, Sablynski C, Burton J, Holtom B (2004) The effects of job embeddedness on organizational citizenship, job performance, volitional absences, and voluntary turnover. Acad Manag J 47:711–722
Lerner RM, Almerigi JB, Theokas C, Lerner JV (2005) Positive youth development. J Early Adolesc 25(1):10–16. doi:10.1177/027243160427321
Machackova H, Dedkova L, Mezulanikova K (2014) Bystander effect in cyber-bullying incidents. Workshop Aggression, Berlin, November, pp 20–22
Mackenzie S, Podsakoff P, Jarvis C (2005) The problem of measurement model misspecification in behavioral and organizational research and some recommended solutions. J Appl Psychol 90:710–730
McAuliffe MD, Hubbard JA, Rubin RM, Morrow MT, Dearing KF (2007) Reactive and proactive aggression: stability of constructs and relations to correlates. J Genet Psychol 167:365–382
McLuhan M (1960) Effects of the improvements of communication media. J Econ Hist 20(4):566–575
Menesini E et al. (2016) Cyber bullying definition among adolescents: a comparison across six european countries. Cyberpsychology. Behav Soc Networking 15:455–463
Menesini E, Nocentini A (2009) Cyber bullying definition and measurement: some critical considerations. Zeitschrift für Psychologie/J Psychol 217(4):230–232
Meyer JP, Allen NJ, Smith C (1993) Commitment to organizations and occupations: extension and test of a three-component conceptualization. J Appl Psychol 78:538–551
Nieman MD (2011) Shocks and turbulence: globalization and the occurrence of civil war. Int Interact 37(3):263–292
Pelucchi S, Sillari G (2016) Intelligence analysis and behavioral sciences. In: Ramirez JM, Fernández JC (eds) Security in infrastructures. Cambridge Scholars Publishing, Cambridge, pp 183–211
Pervin L (1993) Personality: theory and research, 6th edn. John Wiley & Sons, Oxford, England
Piaget J (1972) Intellectual Development. Prentice Hall Inc, New Jersey
Prochaska J, Norcross J, Diclemente C (2008) Changing for good: a revolutionary six-stage program for overcoming bad habits and moving your life positively forward. Amity, Warszawa
Ray RD, Wilhelm FH, Gross JJ (2008) All in the mind's eye? Anger rumination and reappraisal. J Pers Soc Psychol 94:133–145
Rongińska T (2013) Kreatywny charakter współczesnej edukacji. In: Pietrulewicz B, Baron-Polańczyk E, Klimentowska A (eds) Problemy Rozwoju człowieka. Wydawnictwo Naukowe Polskiego Towarzystwa Profesjologicznego, Zielona Góra, pp 19–28
Salzman MB (2001) Globalization, culture, and anxiety: perspectives and predictions from terror management theory. J Soc Distress and the Homeless 10(4):337–352. doi:10.1023/A:1011676025600
Sampson EE (1989) The challenge of social change for psychology: globalization and psychology's theory of the person. Am Psychol 44(6):914–921. doi:10.1037/0003-066X.44.6.914
Smith PK (2007) Why has aggression been thought of as maladaptive? In: Hawley P, Little TD, Rodkin PC (eds) Aggression and adaptation. Lawrence Erlbaum Associates, Mahawah, New Jersey. London, pp 65–84
Spears B, Taddeo C (2014) Model of preventing program keep it tame in Australia, lecture at aggression workshop. Berlin, November 20–22
Small G, Vorgan G (2011) iBrain. Surviving the technological alteration of the modern mind, Polish edn. Vesper, Poznań
Steffen G, Kohl D, Happ D (2014) Predicting workplace mobbing by working conditions, Workshop aggression, Berlin. 20–22 November
Stoll C (2000) Krzemowe medium (The Silicon Medium). Rebis, Poznań
Trempała J (2002) Czas i zmiana. Wydawnictwo AB, Bydgoszcz
Trempała J, Cieciuch J (2016) The analysis of change in behavior and development: on some errors and possibilities to correct them. Curr Issues Person Psychol 4(2):65–74. doi:10.5114/cipp.2016.60168

Tyszka T (2005) Psychologiczne osobliwości pieniądza. Centrum Psychologii Ekonomicznej i Badań Decyzji Wyższa Szkoła Przedsiębiorczości i Zarządzania, Warszawa
Wallace P (2001) The psychology of internet, Polish edn. Rebis, Poznan

Internet Sources

Akmed (2010) The medical consulting centre web site (Centrum Konsultacyjnego Niepublicznego Zakładu Opieki Zdrowotnej). Taken from http://www.akmed.waw.pl/siec. Accessed 11 Aug 2016
Bendyk E (2005) Death and internet (Śmierć i Internet). Taken from http://bendyk.blog.polityka.pl/2007/05/23/smierc-i-internet/. Accessed 22 Sept 2016
Campaign Keep It Tame (2012) Taken from http://www.news.com.au/national/south-australia/online-campaign-aims-to-teach-teenagers-how-to-respect-each-other-on-social-media/story-fndo4dzn-1226515454517. Accessed 11 Aug 2016
Chuderski A (2002) Wykorzystanie metod sztucznej inteligencji w badaniach nad umysłem. Taken 29 Sept 2016 from http://www.kognitywistyka.net/mjkasperski@kognitywistyka.net
Cogito (2011) Groźna bezsenność wśród nastolatków. Cogito. Taken 20 Sept 2016 from http://www.cogito.com.pl/Artykul/38/Grozna_Bezsennosc_Wsrod_Nastolatkow.html
Daily Mail (2013) http://www.dailymail.co.uk/news/article-2516641/4chan-user-sets-live-streamed-suicide-attempt-200-people-watch.html. Accessed 19 Sept 2016
ICD-10-CM (2016) Diagnosis code. http://www.icd10data.com/ICD10CM/Codes/F01-F99/F80-F89/F84-/F84.5 and http://www.icd10data.com/ICD10CM/Codes/F01-F99/F20-F29/F20-/F20.1. Accessed 29 Sept 2016
Mobirank (2016) Taken from https://mobirank.pl/2015/09/12/czas-spedzany-w-internecie-na-swiecie/. Accessed 11 Aug 2016
Taler H (2014) The mass histery. Taken from http://www.spidersweb.pl/2014/11/masowe-histerie.html. Accessed 11 Aug 2016

Some of Broken Heart Pages

www.lifehack.org/articles/communication/how-heal-broken-heart-and-the-science-behind-2.html
www.wikihow.com/Heal-a-Broken-Heart
www.huffingtonpost.com/joyce-marter-/broken-heart_b_4645774.html

Author Biography

Dr. Marzanna Farnicka is a licensed psychologist and adjunct professor in the Institute of Psychology at the University of Zielona Góra. She is a trainer of social skills and ART therapy. She is a member and chair of local branch of the Polish Psychology Association, the International Society for Research on Aggression and a cofounder of the Research Group on Family and Adolescents in Middle Europe. Her research includes developmental psychopathology included issues connected with family life, coping with stress and aggression and the support of development. She chairs the Polish CICA.

Part II
Cybersecurity

Cyberspace: A Platform for Organized Crime

Natividad Carpintero-Santamaría and María Pilar Otero

Abstract Organized crime is a multi-billion dollar business and is growing in scale. Cyberspace has become a very powerful platform that facilitates the protection of the criminal economy and its activities by means of complex technological and financial schemes. New concepts such as cybercrime and cyber money laundering are socially destabilizing practices that constitute a challenge that must be combatted. Organized crime organizations act in a transnational context, increasingly moving in large groups that have proven in the 21st century to be perfectly capable of adapting their activities to the globalized context. The cyberspace platform has contributed to an expansion of organized crime that facilitates their activities and makes it more difficult to identify the wide scope of operations that fit all kinds of crimes. Criminals move in cyberspace with an impunity that was unthinkable a few years ago. This chapter will focus on the analysis of the following issues: 1. Cyberspace as a key instrument for the spreading of criminal activities, and legal mechanisms to combat them. 2. The complexities involved in both a sociological and a legal definition of organized crime. 3. Money laundering and criminal legal instruments to combat practices such as confiscation. 4. Illegal trafficking of dual-use materials with a specific focus on illicit trafficking of chemical, biological, radiological and nuclear (CBRN) materials and the risk posed in proliferation.

Keywords Cyberspace · CBRN illicit trafficking · Organized crime · Money laundering

Submitted: 26.8.16; Accepted: 16.10.16.

N. Carpintero-Santamaría (✉)
Instituto de Fusión Nuclear, Universidad Politécnica de Madrid, Madrid, Spain
e-mail: natividad.csantamaria@upm.es

M.P. Otero
Departamento de Derecho Penal, Procesal e Historia del Derecho, Universidad Carlos III de Madrid, Madrid, Spain
e-mail: potero@der-pu.uc3m.es

J.M. Ramírez and L.A. García-Segura (eds.), *Cyberspace*, Advanced Sciences and Technologies for Security Applications, DOI 10.1007/978-3-319-54975-0_7

1 Cyberspace: A Haven for Organized Crime

As is known, the internet is the most common form of personal, work, and social communication with the particularity that it is a network of networks that interconnects millions of computers, knowing no borders and without any governance. Unfortunately, this parallel technological development has generated a dark side: it allows the use of computer networks to commit new crimes (illegal access to computer systems, data erasure, etc.) and facilitates the performance of traditional crimes (such as scams, money laundering, terrorism, or the sexual harassment of minors). Cybercrime is characterized by macro-victimization, that is, the distribution of the effects of crime at breakneck speed and an unlimited number of users, with the added problems of the difficulty in the pursuit since it is a transnational crime.

The internet favors anonymity. Although it is said that anonymity is no longer a feature of internet since it is increasingly easier to identify the IP addresses, the truth is that today it has become more complex, despite the digital traces left by crime, to identify the authors of these behaviors, in comparison to those other subjects who commit similar offenses in the real world. Criminological theories assert that, if an action is perceived to be executed anonymously, there will be a resulting increase in the sensation of impunity which, in turn, triggers an increase in the risk of the agent actually carrying out the offense.

In short, cyberspace is a haven for organized crime (Miró 2012, p. 143 et seq.). This makes it necessary, firstly, to review the mechanisms of international cooperation; second, to review the principles of criminal law enforcement, replacing territorial jurisdiction based on the principle of sovereignty by the principle of ubiquity; third, to specialize in criminal investigation (Flores Prada 2012, p. 309 et seq.); and, fourthly, to harmonize criminal laws regarding cybercrime.

The reforms of the Spanish Penal Code, affected by both LO 1/2015 and 5/2010, made a significant effort to adapt the domestic criminal law to European regulations in the field of organized cybercrime. Some significant changes are the following:

1. *Punishment of child pornography.* The Spanish Penal Code defines child pornography, as stated in Directive 2011/93/EU, on combating sexual abuse and sexual exploitation of children and child pornography, which covers not only material depicting a minor or disabled person engaging in sexual behavior, but also realistic images of minors engaged in sexually explicit conduct, even though they do not reflect real facts. Where child pornography is concerned, acts of production and distribution are punished, as are those who knowingly attend pornographic performances involving minors or people with disabilities in need of special protection. The mere use or acquisition of child pornography is also punishable, and a new paragraph is included to punish those who acquire this type of pornography through information and communications technology, on the understanding that the new technologies are a major gateway to providing support for pornography. For this reason, judges and courts are expressly empowered to adopt measures either to shutdown internet pages containing or

disseminating child pornography or, if necessary, to block access to those pages. Protection of minors from abuse over the internet or other means of telecommunication, due to the easy access and the anonymity they provide, is reinforced with a new paragraph in Article 183b of the Criminal Code intended to punish those that try to contact a fifteen-year-old minor, or performs acts aimed at duping minors into providing pornographic images (child grooming).

2. *Punishment of illegal access to a computer system* (*hacking*) is introduced by the LO 5/2010 in compliance with DM 2005/222/JAI of the European Union Council on crime against computer systems, and reformed by the LO 1/2015 transposing Directive 2013/40/EU of August 12th, on attacks against information systems and electronic data interception when they do not constitute a personal communication. According to the abovementioned Directive, a clear separation is introduced between the alleged disclosure of data that directly affect personal privacy and access to other data or information that may affect privacy but are not directly related to personal privacy. For example, it is not the same to access a list of contacts as to collect data through a software program. The Directive also includes: (1) A definition of interception of transmissions between systems when they are not personal transmissions to typify automatic (non-personal) transmissions between computers. (2) Punishment of the facilitation of software produced, or specifically designed or adapted, for the committing of these crimes. An aggravated penalty is imposed for the above crimes when they are committed within the framework of a criminal organization or a criminal group.
3. *Regulation of alleged cases of computer damage (attacks on data, computer programs, and foreign electronic documents) and illegal interference with information systems.* In compliance with the aforementioned Directive, these cases are regulated separately. With respect to computer damage, two main types are taken into consideration. 1. Crimes committed within the framework of a criminal organization. 2. Attacks on the computer system of a critical infrastructure or a system that can imply serious danger to state security, the European Union, or a Member State of the European Union. The Criminal Code defines critical infrastructure as "an element, system, or part thereof that is essential for the maintenance of vital societal functions, health, safety, security, and economic and social welfare of the population, the disruption or destruction of which would have a significant impact that would prevent it from maintaining its functions" (Art. 264.2. 4ª). In this regard, Law 8/2011 of July 17th, on measures for the protection of critical infrastructure points out: "For the purposes of this Act, critical infrastructures are strategic infrastructures whose operation is indispensable and does not allow alternative solutions, so that its disruption or destruction would have a serious impact on essential services" (Art. 2). This Act calls for the development of a national catalog of strategic infrastructures, which would be secret.

   Whenever the illegal jamming of information systems takes place, the interference or interruption must be deemed severe. One of the most important means of achieving this interference is through the transmission or introduction of new data; interventions that are known as DDoS (Denial of Service) attacks. In such cases, overload due to saturation of the system (as occurs, for example, when there is a massive use of spam) produces a disablement of services, remaining inoperative for a while and suffering consequent damages.

4. *United Nations Security Council Resolution 2178, adopted on September 24, 2014, modifies terrorist offenses.* It is noteworthy that international Jihadist terrorism has incorporated new forms of aggression, consisting of new instruments for recruitment, training or indoctrination through the internet and especially through social networks, with messages designed to provoke terror in the population, accompanied by calls to commit attacks.

   To combat these new threats, as well as the traditional behaviors of criminal organizations or terrorist groups, criminal reform updates the regulations, firstly, to accommodate the phenomenon of individual terrorism and, secondly, to typify indoctrination and military training or handling of any kinds of weapons and explosives, with special mention of what is done through the internet or services accessible to public communication.

2 Sociological and Legal Definition of Organized Crime in the European Union (EU) and the United Nations (UN)

Organized crime has become more sophisticated in the globalized business environment. It works within a digital world that makes it much more difficult for law enforcement to intervene (Otero 2011, p. 110 et seq.). The European Security Strategy adopted in December 2003 identified it as one of the main global challenges and a key threat to security. And five years later, the Report on the Implementation of the European Security Strategy (2016) identified terrorism and organized crime as the second greatest threats to the European Union (EU) security.

The definition of *organized crime* can be analyzed from three different perspectives (Flores 2007, pp. 141–191): organizational/structural, relational (*vis-a-vis* clients and the socio-political environment), and a profit business model that involves an underground market. Structurally, criminal organizations are formed through the use of violence, corruption, and the obstruction of justice (Herrán et al. 2007, pp. 21–26). These organizations use violence as a means to establish control over their own members, thus guaranteeing internal discipline. They assert themselves over competitors in the illicit markets that they intend to control, revolving around drugs, prostitution, people smuggling, arms trafficking, money laundering, etc. Organized crime groups use corruption, including bribes, to expand their activities. In several cases, this corruption has penetrated legitimate criminal justice institutions, and police circles, in particular, obstructing justice and arousing subsequent mistrust among citizens as to the legitimacy of the rule of law.

In recent years, a new characteristic has emerged: a development that involves the internationalization of organized crime. The economic, political, social, and technological changes that have occurred since the last decade of the 20th century, have diffused national borders, causing a change in the modus operandi (MO) of organized crime. This new MO goes beyond the political barriers of each State, creating pervasive transnational illicit markets.

The power of a criminal group lies in its real capacity to threaten with violence, rather than in its effectiveness at carrying out crimes (Mapelli 2001, p. 2088). The part of the definition that makes reference to the accumulation of political or economic power is fundamental as it alludes not so much to a reality, such as a particularly serious crime committed by its members, but rather to the potential danger that emanates from that accumulation of power, given its capacity to accumulate economic resources (Moore 1986, pp. 50–54).

The EU, through international agreements signed primarily after 1998, established a long list of conditions and requirements necessary for a criminal organization to be classified as an organized crime group and its illegal behavior categorized as organized crime. These conditions and requirements are: (1) the organization has existed for a certain time; (2) members show convincing evidence of criminality, and, (3) its objective must be the accumulation of political and/or economic power. Additional criteria were established by the EU: (1) the existence of a certain discipline and internal control within the organization, and (2) the use of violence as a means of committing offenses (Fernández Steinko 2008, p. 23).

Specifically, the EU has promulgated the following three instruments:

(I) *The Joint Action (98/733/JHA)*, adopted by the Council based on article K.3 of the EU Treaty (21/XII/1998), makes it a criminal offense to participate in a criminal organization in any EU Member State.

Article 1 establishes: "With the meaning of this Joint Action, a criminal organization shall mean a structured association, established over a period of time, of more than two persons, acting in concert with a view to committing offenses which are punishable by deprivation of liberty or a detention order of a maximum of at least four years or a more serious penalty, whether such offenses are an end in themselves or a means of obtaining material benefits and, where appropriate, of improperly influencing the operation of public authorities".

Article 2 stipulates the obligation on Member States to declare as a criminal offense one or both of the following activities: "to participate actively in the criminal activities of the organization or merely to agree that the criminal activity should be pursued."

(II) *The Council Framework Decision 2008/841/JHA* revokes the Joint Action instrument regarding the fight against organized crime.

Article 1 defines a criminal organization as "a structured association, established over a period of time, of more than two persons acting in concert with a view to committing offenses which are punishable by deprivation of liberty or a detention order of a maximum of at least four years or a more serious penalty, to obtain, directly or indirectly, a financial or other material benefit".

Article 2 stipulates that each Member State regards as an offense one or both of the following: (a) participation in the criminal activities of the organization, including financing or agreeing to pursue the criminal activity without necessarily doing so; and (b) taking part in the actual execution of the activity.

The first solid example of the United Nations' involvement in this head-on battle against organized crime is the promulgation of the UN Convention on Transnational Organized Crime (UNCTOC) (AKA: Palermo Convention), signed in Palermo on November 15, 2000 (General Assembly Resolution A/RES/55/25). The notion of organized crime is specified in its Article 2 as "a structured group of three or more persons, existing for a period of time and acting in concert with the aim of committing one or more serious crimes or offenses established in accordance with this Convention, in order to obtain, directly, or indirectly, a financial or other material benefit".

Article 5.1 also establishes that each State Party shall establish as a criminal offense: (a) agreement to commit a serious crime, with or without the need for the conspirators to have participated actively in the eventual execution, active participation in the illicit activities of the criminal organization, or active participation in the other activities of the organization, so as to contribute to its criminal aims; and (b) the organization, direction, aiding and abetting of the facilitation or counseling of the commission of a serious crime involving an organized criminal group. This line of thought was initiated by German criminal jurisprudence that defines organized criminality as "the planned perpetration of crimes induced by a desire to obtain profits or power, an aspiration that can come to be very important when two or more participants divide the work during an undetermined period of time: (a) using semi-managerial or semi-professional structures, (b) employing violence or other intimidating mechanisms, and (c) influencing politics, the media, the administration, justice or economy" (Kinzig 2004, p. 57).

(III) Resolution of October 23 2013 on organized crime, corruption, and money laundering: recommendations on action and initiatives to be taken by the European Parliament (Final Report (2013/2107(INI)).

In the majority of European legislation, based on the UNCTOC definition of organized crime, acts of terrorism are explicitly excluded from organized crime offenses because these are considered a particular and specific kind of crime (Fernández Steinko 2008, p. 23). The 2010 Conference of Ministers of Justice of Ibero-American Countries (COMJIB), at its 17th Plenary Session, adopted an intermediate position on the definition of organized criminal behavior when defining illicit associations without regard to the aim of obtaining material benefit: "Structured groups of at least three persons that exist for a permanent or temporary period of time with the aim of committing crimes" (COMJIB 2010, n.p.). And it adds as well that "only in the case of disregarding irrelevant facts or with little criminal significance is there the possibility of foregoing punishment for associated criminal behavior referring to crimes considered to be minor" (COMJIB 2010).

Finally, in its art., 571 bis, the last Spanish Criminal Code reform (*Ley Orgánica* 5/2010, 22nd June) introduced for the first time a definition of organized crime, even if it does not include the aim of obtaining material benefit: "For the purposes of this Code, a criminal organization is understood to be a stable group formed by more than two persons existing for an indefinite period of time which divides tasks or functions in a coordinated manner with the aim of committing crimes or repeatedly committing minor crimes".

These changes in the definition of organized crime over time demonstrate that the criminal law concept of organized crime has progressively broadened. The Palermo Convention defines organized crime either as conspiring to commit a crime or as active participation—actually committing a serious criminal offense; it does not place a limitation on a particular deprivation of liberty sentence and focuses on obtaining a material benefit. The COMJIB definition excludes the concept of minor crimes and defines the criminal organization as being either permanent or temporary. And the last reform carried out through the Spanish Criminal Law 5/2010 extended the definition of organized crime to include conspiring to commit minor crimes.

Evidently, the fewer requirements needed to constitute the concept of organized crime, the smoother the road to criminal prosecution, since the burden of proof is reduced. However, the unlimited broadening of the concept will generate new problems of legal insecurity and the risk of leaving interpretation in the hands of the judges. Finally, it will be necessary to analyze whether the broad scope of offense classification is coherent and therefore, justifiable in relation to the established right that it intends to protect. A priori, the broadening of the concept to include minor crimes, apart from distorting the essence of organized crime, demonstrates an attack on the principle of proportionality.

In short, the main differences between the European legal framework (2008 Council Framework Decision), the 2000 UN Convention, and the Spanish definition of the offense can be found in three specific areas (Otero 2013, p. 107 et seq.).

First, the models of criminal organization in the European legal framework and the UN Convention include organized crime induced "by material benefit" (except in the 1998 Joint Action whose article 1 establishes that the offenses can constitute an end in and of themselves or be a mean to obtain material benefit and, in this case, improperly influence public authority). This means that a complex structure, such as that of an organized criminal or terrorist group, supported by a lucrative aim, must be added to the potential danger of destabilizing the social order and generating a loss of confidence in the rule of law (Brandaríz 2009, p. 745). According to Brandaríz, this greater harm, which goes beyond the infringement of established rights, specifically affects the stability of current economic and political order. Therefore, a specific punitive policy is justified. However, this lucrative aim was not targeted by the Spanish legislation. The 4th point of the 2008 Council Framework Decision, derived from Article 2a, establishes the freedom of Member States to classify as criminal organizations other groups whose aim is not to obtain financial gain or other material benefit, therefore going beyond the obligations stated in the Article. This inclusion of limiting criteria increases the burden of proof and consequently constitutes a stumbling block for criminal prosecution (Zúñiga 2002, p. 280; Brandaríz 2009, p. 753). In this context, the COMJIB, in its 17th Plenary Session (2010), accepted that the complexity and specialization of criminal organizations make it inappropriate to include a criterion such as lucrative aim because it poses a restriction to the concept of "unlawful assembly". Under this understanding, whatever may be the aim pursued, forming a group to commit any offense of any type should be considered in and of itself criminal.

Second, only serious crimes, that is, those punishable by deprivation of liberty or a detention order of a maximum of at least four years, are classified as offenses within the context of the EU. However, in the Palermo Convention, "serious crimes" are alluded to generically without specification, although the phrase "offenses established in accordance with this Convention" is added. This means a broadening of the scope of the offense, since the crime, in the latter case, does not have a threshold of gravity.

Third, the Spanish reform opts for a much greater penalty than that provided in the 2008 Council Framework Decision. Whereas the latter stipulates a maximum of five years imprisonment (Article 3.1b), the Spanish Criminal Code establishes that, in certain circumstances, a maximum of twelve years imprisonment may be imposed (Article 570, bis, 2nd last subsection).

Another important change in the new Spanish legislation is stated in the preamble to the LO 5/2010. This change is centered on giving a criminal law answer not only to criminal organizations (for example, those involved in drug trafficking), which requires proof of a *permanent* structure but also to other similar phenomena present in society today, such as terrorist groups. Although these groups do not meet the structural requirements of LO 5/2010 for a criminal organization, they can, and do at times, engage in extreme violence. Accordingly, the legal reform has responded to this reality by defining, in parallel with organizations, what are known as *criminal groups,* precisely by exclusion, that is to say, as forms of criminal agreement that do not fit into the archetype of the cited organizations, but which do contribute an added criminal danger to the actions of their members (Article 570 ter). Therefore, when LO 5/2010 allows EU Member States to be free to classify other groups of persons as criminal organizations, it seems to invoke the 4th point of the 2008 Council Framework Decision again, beyond the obligations stated in its Article 2a.

The structure of the new offenses corresponds to a similar model in both cases: organizations and groups. However, on the one hand, punishment is stiffer when applied to organizations whose more complex structure corresponds to a deliberate goal of posing a threat to security and legal order. On the other hand, the punishments' distinct nature demands some differences in the description of the actions that are characteristic of the offense at hand.

3 Money Laundering

Money laundering is a process whereby large amounts of money from illegal sources such as trafficking in drugs, people, weapons, terrorist activities, corruption, fraud, and other criminal activities are disguised as legal provenance. Once money has been washed, one of the objectives of this practice is to obtain yet more economic benefits. Through this practice, criminal organizations intend to infiltrate established societies at a legal level. A historically famous example of money laundering involved Al Capone, one of the most famous Chicago gangsters of the 1920s and 30s, who pioneered investment of money obtained from mafia operations in pizzerias, cafes and casinos. Cuesta Sauquillo points out: "Money laundering, required for all criminal organizations, consists of hiding, processing and integration of the assets of criminal origin in the legal economic system. Money laundering can be treated from three perspectives: (1) administrative or preventative; (2) judicial or criminal and (3) investigative or police" (2012, p. 246).

Investments in houses, real estate, land, luxury cars, gold, antiques, and so forth are common in the practice of money laundering anywhere in the world. The current process of globalization at commercial, economic and social levels has boosted the performance of numerous entities, based on complex financial engineering practices and schemes that favor the concealment of illegal origin of certain capitals. The lifting of trade barriers as an essential element of the process of globalization has also led to the liberalization of financial markets and the free movement of capital. In addition, this policy has opened avenues of illicit or illegal practice that leverages the current complex financial mechanisms for money laundering. For their activities, money launderers seek out countries where lax judicial frameworks are a given, or there is not an effective anti-money laundering policy. Financial engineering practice, as well as the expansion of communications over the internet, contributes to money laundering and facilitates anonymity and impersonation. This provides rapid technological advances to hide and hinder information. Money laundering affects not only the economies but also the political stability of countries; one of its worst consequences being the continuation of criminal activities within a vicious circle. According to the UN Office on Drugs and Crime (UNODC): "Transnational organized crime is estimated to generate $870 billion a year—more than six times the amount of official development assistance and the close to 7% of the world's exports of merchandises" (2009, p. 1).

The case of money laundering, occurring in both organized criminal and terrorist groups, constitutes a clear example that criminal law functions from a prospective (in this case the reference point is a future act) instead of a retrospective (focusing on the committed crime) point of view. It is a legal instrument that does not only or so much judge what *has* happened, but rather what *could* happen in the event that illicit money was effectively used to infiltrate the legal economy and society (Fernández Steinko 2007, pp. 96–97; Bajo and Bacigalupo 2009, pp. 11–252).

At the same time, money laundering embraces the notion of the degree of potential danger posed by organized crime. Even if there is no evidence that a person has committed a violent crime, s/he could be arrested for having attempted to launder money presumably obtained from non-violent crime and for everything that s/he could do with this money. The laundering is, therefore, separated from the main crime. It is even a different offense if the person is unable to demonstrate the origin of his/her assets. The individual could even be punished more severely than for the potential violent crime, which might be seen as a breaching of the principle of proportionality of punishment. At the same time, it also represents a reversal of the burden of proof, violating the principle of presumed innocence, since it is the accused party that has to demonstrate the non-illicit origin of the assets under suspicion.

Along these lines, the 2010 Spanish criminal legislation, according to the European Union (Directive (EU) 2015/849 of the European Parliament and of the European Council of May 20th 2015 on the prevention of the use of the financial system for money laundering or terrorist financing) through the modification of articles 301 and 302 of its Criminal Code, broadens the criminal classification of money laundering to include two new features: first, the utilization and possession of assets known to come from criminal activity, even if the crime was committed by the very subject that possesses the assets; and, second, punishing the conduct of self-laundering. This double punishment may violate the principle of *non bis in idem* (Manjón Cabeza 2010, pp. 344–345).

A report from the General Council of the Spanish Judiciary (CGPJ Report 2008, p. 118) suggests that the traditional classifications of money laundering have in common the characteristic that they tend to dissimulate the illegal origin of the assets, giving an appearance of legality to the proceeds or profits of the crime. So, concealment, complicity, transmission, and acquisition mean an apparent change of ownership that situates the property within the assets of another person who has not committed the crime, with the purpose of incorporating it into the legal and economic traffic. However, when the 2010 law sanctions those who simply possess or utilize property known to be the product of criminal activity (or even of gross negligence), it is not categorized as incriminating conduct that actually constitutes money laundering, since this conduct does not involve a real or apparent change of ownership.

The 2008 report of the Spanish *Consejo Fiscal* (pp. 162–163) (a board that assists the Chief State Prosecutors in their duties) suggested using the term "criminal activity" to describe money laundering, rather than the term "offense." The *Consejo Fiscal* reasoned that since a prior conviction on the predicate offense was not necessary, in accordance with the Supreme Court, the guarantees associated with evidence from the investigation were relaxed. The introduction of this nuance is coherent with the terminology used in the new classification of confiscation (a category closely related to the offense of money laundering). Confiscation is characteristically carried out in response to organized crime, where it is not always possible to prove the concrete participation of a particular member in a particular criminal activity (due to the interchangeability of members), although his or her continuing membership in the organization can be proven, which in and of itself constitutes "criminal activity".

This new category of crime in the fight against money laundering makes it clear that the designation is not so much an instrument to facilitate the pursuit of the crime, but is, rather, primarily useful in dismantling the economic structural power in the world derived from organized crime (Fernández Steinko 2008, p. 116) and terrorist activity.

3.1 Tax Havens and Bank Secrecy

Tax havens have existed throughout history, from the time of the corsair pirates who set up the center of their economies in the Caribbean islands. Bank secrecy, confidentiality, and customer facilities to establish their businesses are the main characteristics of tax havens.

Banking secrecy is a special law that allows banks to protect customer financial data, personal data, deposits, account numbers, transactions, and the like, against third parties. In principle, this secrecy can only be lifted by a court order, although in recent years, due to international pressure, some tax havens have relaxed this traditional practice. It has allowed and still allows the development of activities such as money laundering and other fraudulent activities, although money laundering can follow its course without banking secrets. The lifting of bank secrecy remains a complex issue, as it is subject to the requirements of the banks involved. In order to proceed with the lifting of banking secrecy, banks ask to be provided with certain information, if there is a suspicion that a crime has been committed, or that information requested cannot be obtained by other means, or the government requesting it can provide very specific information from the person they are investigating, such as their account number.

Money laundering can be conducted by several procedures, and cyberspace provides numerous advertisements of financial groups that facilitate the opening of current accounts in offshore jurisdictions. Accordingly, these legal instruments can be used for money laundering. Offshore societies are companies created for lawful purposes offering tax advantages. They are located outside the money owner's country and in several cases these societies provide enough margin for money laundering purposes, as they have no physical presence and no employees, just a mailing address.

Despite the role of these societies, which have become an integral part of the economy of globalization, their use as a means to launder money is a circumstance that worries institutions and states. Shell companies in offshore jurisdictions are the most difficult to track and investigate, because in these jurisdictions there are mechanisms that aid concealment of owner profits from procedures such as the issuing of shares under different certificates, thereby hiding the identity of the true owners.

Offshore financial centers are areas in which business can be carried out by non-residents, permitting the legal creation of impenetrable business networks, and allowing financial transactions to be carried out by businesses with third parties in foreign countries without having to notify any authority. These financial centers and jurisdictions are created, among other reasons, to ensure their clients' confidentiality; they establish companies, intermediaries, or trusts designed to make it impossible to identify the real owners or final beneficiaries of the accounts; and they use computing and accounting techniques to make said accounts disappear (Fernández Steinko 2008, pp. 183–194). The aforementioned Communication of the Commission to the European Parliament and the Council (Brussels 20-11-2008) recommended inclusion of the provisions of the 2001 Protocol, which established that the authorities of the Member States should facilitate information about the accounts and bank operations of certain identified persons and that the banking secret cannot be cited to avoid cooperating in this context.

3.2 Criminal Legal Instruments to Combat Money Laundering. Confiscation and the Financial Action Task Force (FATF)

A few decades ago, the term "Gross Criminal Product" was used to describe the money moved by drugs, illegal arms sales, coerced prostitution, urban development corruption, and people trafficking. Contrary to what one might think, this parallel economy based on crime is not fed primarily through blue-collar crime.[1] The primary actors, as well as main beneficiaries, are groups infinitely more powerful than all the blue-collar criminals put together (Fernández Steinko 2008, pp. 276–285). Given their organizational, planning capacity, and leadership within the organized criminal groups, they hold some of the great fortunes of the world, run large multinational companies and, more concretely, engage in tax evasion, avoiding democratic control of their activities and operating outside the parallel institutional structures of the most influential states in the world.

Along with the crime of money laundering, another instrument to isolate the offender integrated into organized crime, and thus render offenses unprofitable, is the confiscation of profits. As González Mas points out, "the treatment of both figures (confiscation and money laundering) is of particular interest to combat organized crime. Indeed, this type of crime is characterized by using technologically advanced resources that can invest large amounts money in order to ensure the effectiveness of its activities which it is always very lucrative" (2012, p. 157).

Indeed, the danger of organized crime is defined through its capacity to accumulate profits that can be used to infiltrate the economy. Since the 1980s, the fight against organized crime has been focused, to a great extent, on tracking associated

[1]According to Jose Rivera, "blue-collar crime refers to crimes that are somewhat more obvious and easily detected by police authorities. Blue-collar crime is often associated with geographical regions with low income or with over-population issues". See: http://www.legalmatch.com/law-library/article/what-is-blue-collar-crime.html#sthash.Xi37UC5.dpuf.

finances in order to broaden the scale of crime, to include corrupt civil servants, or to assume a monopoly of violence (Fernández-Steinko 2008, pp. 92–96). Therefore, the essential tool in this battle to weaken the economic power of the organized criminal group is confiscation.

The European provisions over the last fifteen years have been directed at two issues related to this discussion: first, insuring the harmonization of national provisions related to confiscation (and penal sanctions applicable to money laundering), and second, applying the principle of mutual recognition to resolutions for the preventative embargo of property, for securing evidence between different countries, for confiscation, and for the weight of the burden of proof to be met regarding the origin of the property affected.

Directive 2014/42/EU of the European Parliament and of the Council of April 3rd 2014 on the freezing and confiscation of instrumentalities and proceeds of crime in the European Union recognizes that confiscation, as well as the recovery of proceeds of criminal origin, constitutes a very effective tool to fight organized crime that is essentially motivated by the desire for profit. In effect, confiscation impedes the use of ill-gotten funds for financing other illegal activities, a practice that undermines confidence in the financial systems and corrupts legitimate business.

A case could be brought before a civil court grounded in the presumption that certain properties come from criminal activities. In these cases, the burden of proof is reversed, having the accused to prove the legal origin of the property. With the burden of proof lying in the hands of the accused, this remedy would serve as an easier and speedier intervention than would a criminal trial.

The creation of an offense of possession of ill-gotten property has been proposed within the European law framework to pursue the proceeds of crime in those cases in which the value is disproportionate to the income reported by the owner. In this case, the proceedings are carried out before a criminal court and the burden of proof is not completely reversed (this classification exists in French criminal law).

The UN Convention on Transnational Organized Crime of 15/XI/2000 already proposed an equivalent confiscation law (Battista 2007) that reduces the investigative burden when it is not possible to determine the price or the profit of the crime. This law can use other property of similar value that the accused has at his or her disposal to determine the criminally obtained property's monetary worth.

The reform of the Spanish criminal legislation (LO 5/2010), through modification of article 127, complies with the Council Framework Decision 205/212/JHA of February 24th, 2005, regarding confiscation of products, instruments, and proceeds from crime. The preamble of the law highlights that the main objective of organized crime is profit and, consequently, establishing common provisions for monitoring, embargo, seizure, and confiscation of the proceeds from crime is the priority for carrying out an effective fight against this criminal activity.

As it relates to organized crime, the existing provision on confiscation was completed by the Council Framework Decision 2002/475/JHA and applies to the fight against terrorism. The provisions enable the judiciary and the courts to come to an agreement as to the effects, property, instruments, and profits obtained from criminal activities committed by a criminal or terrorist organization or group. To facilitate the measure, it is presumed that criminal activity has occurred when the value of property proceeds is disproportionate to the legal income of each person convicted of crimes committed within a criminal organization or group. Furthermore, judges and courts are empowered to approve confiscation when it is a negligent offense that carries a prison sentence greater than one year.

LO 1/2015 introduced an ambitious revision that presents important modifications intended to provide legal instruments that are more effective in the recovery of assets from crime and in the economical management of them. The reform takes into account the European Directive 2014/42/EU of April 3, on the seizure and confiscation of instrumentalities and the proceeds of crime in the European Union.

In 1989, the G7 countries (Germany, USA, Canada, France, UK, Japan, and Italy) promoted the FATF, which is a supranational body to promote effective implementation of legal and operational ways to combat money laundering practices. Present priorities of the FATF are counter terrorist financing; enhancing transparency; providing outreach to the FinTech community; effective implementation of FATF standards; increased operational focus and enhanced international standing. FATF observer entities including Europol, Interpol, the European Central Bank, the International Monetary Fund, and the World Bank, among others.

The Spanish *Servicio Ejecutivo de Prevención de Blanqueo de Capitales* (SEPBLAC) is a part of FATF and works as a multidisciplinary financing intelligence instrument under the Bank of Spain, the State Tax Administration Agency, and the National Police and Guardia Civil. SEPBLAC's purpose is to make the fight against money laundering increasingly effective. Spain is developing the Financial Ownership File, similar to those developed in Germany and France. The purpose of these databases is to the streamline processes of identification and confiscation of assets derived from criminal activities. The procedure involves automatic access to the account data of individuals suspected of carrying out money laundering activities, and its objective is to prevent and deter money laundering and terrorist financing, based on information obtained in current accounts, security accounts and deposits, etc.

The flow of criminal money and tax evasion generates a high degree of social, political, and economic harm. This fact should lead to a search for further possible solutions that help to halt money laundering and its associated problems. The creation of independent international regulatory agencies to monitor international financial activity would assist in the fight against these types of parallel financial structures. Aside from the further creation of legal instruments to fight money laundering, it should not be forgotten that the structural power derived from organized crime is sustained primarily because of the role played by tax havens, that is to say, offshore financial centers and jurisdictions with bank secrecy.

In addition to the legal tools described above, computing technology applied to combat terrorist crimes, together with intelligence services, constitute some of the most important instruments for unraveling the complexity posed by the activities of these organizations (Herrán et al. 2007, pp. 67–72). Therefore, workers in the judicial system must be empowered to utilize technology and intelligence in their investigations. In addition, international cooperation must be emphasized, given the transnational character of organized crime and terrorist activity.

4 Illicit Trafficking of Dual Use Materials: The Case of Chemical, Biological, Radiological, and Nuclear (CBRN) Materials

Combating illicit trafficking of dual-use materials is an arduous task due to the opacity of these camouflaged operations; indirect transmission; diversification of supplier countries and, in some cases, the participation of states themselves. The development of new technologies for transport and communication of goods via cyberspace, not only substantially facilitates the flow of illegal trade, but also enhances security for traffickers.

One of the most worrying concerns about the illicit trafficking of dual-use material is related to the destiny of chemical, biological, radiological, and nuclear materials (CBRN), because of the proliferation problem they might trigger.

Covert acquisition of dual-use materials through illicit trafficking is produced by applying a series of techniques and strategies that are becoming increasingly sophisticated. Traffickers will falsify a product by placing misleading labels and inventing a destination it will never reach. A key factor in illicit trafficking is the falsification of the concept for which CBRN materials will be used: strains for the preparation of vaccines that could be used for bioterrorism, or radioisotopes for application in medicine, which could also be used to make a dirty bomb.

Illegal trafficking of dual-use materials is usually in the hands of traffickers equipped with solid structures in the shape of legal defense or front companies[2] that protect their activities. It is possible as well, to find opportunistic or corrupt officials acting as traffickers who allow bribery and turn a blind eye to certain crimes in exchange for extra profit.

The illegal trafficking system today allows manufacturers to purchase raw materials in one country, buy components in another country, assemble the product in another, and then send the finished product to a different destination country. This operational system is known as *triangulation*. Traffickers may also falsify the end user or work through front companies to purchase a product on behalf of others. The latter technique shows how smuggling networks are evolving in response to the challenges posed by strict export controls of such materials. Sometimes, organized offenders are businessmen, and their organization acts as a criminal enterprise, assuming their own models, industrial structures, and business. Another technique is the use of intermediaries and people who will acquire the materials or send

[2]A front company is a subsidiary or shell company used to shield another company from liability or scrutiny. A front company can be used to protect a parent corporation or brand from negative publicity in the event of a mishap, and may also be used to conceal illegal activities.

people to seek the material, along with the networking activities that will coordinate the shipments. In addition, CBRN materials trafficking may occur by transporting them through countries where there is no strong demand for exports control. Traffickers will also resort to practices such as bribery or corruption in economically depressed areas.

Another way of illegally trafficking with CBRN agents or materials is to transfer intangible technology and know-how. Manuals for making weapons with CBRN agents are found in cyberspace. Cyber dissemination of these manuals becomes a rather difficult task to combat as they proliferate at several internet levels.

Open acquisition of CBRN materials, based on legal trade, uses commercial and regulatory channels that reflect user data, destination, and the end use of the legally acquired material. Open procurement of dual-use materials is subject to a number of regulations. In Spain, dual-use exports, especially those products associated with defense material transfers at global, intercommunity, and individual levels, are subject to comprehensive control by the Regulatory Inter-Ministerial Board of Foreign Trade on Defense and Dual-Use Material (JIMDDU), created in 1993; and the Special Register of Foreign Trade Operators in Defense and Dual-Use Material. The JIMDDU carries out its activities under the Ministry of Industry, Tourism, and Trade and has representatives from the Ministries of Industry, Tourism and Trade, Foreign Affairs and Cooperation, Interior, Defense, and Treasury. Also in Spain, the Law 53/2007 of December 28th controls the foreign trade of defense material and dual-use. This Act was extended by the Royal Decree 2061/2008 of 12 December. The Organic Law 12/95 of December 12th on Repression of Contraband was amended by Organic Law 6/2011 of June 30th, acting on criminal matters.

The Spanish governmental instruments to control dual-use materials work in compliance with the EU Regulation (EC) Nº 428/2009, which establishes that under the EU regime, the export of dual-use items may not leave EU customs territory without an export authorization. With respect to the possible clandestine export of items in connection with CBRN weapons or ballistic missile programs, the EU Regulation includes a "catch-all clause", as well as certain restrictive measures. These measures are being applied to thetrade of dual-use items with Democratic People's Republic of Korea, Iran, and Syria (The EU Dual-Use Exports Control Regime. European Commission). Resolution 1540/2005 of the United Nations also requires member countries to take measures to address the participation of non-state actors in the proliferation of weapons of mass destruction (WMD) infrastructure.

To reinforce the prevention of WMD proliferation and their means of delivery, a number of export control regimes have also been established, namely: the Australia Group; Missile Technology Control Regime (MTCR); Nuclear Suppliers Group (NSG); and the Wassenaar Arrangement and Zangeer Committee.

Despite all the legislation in force, both at national and international legal level, there are still vulnerabilities in the system. As Moran points out:

> Although today we have a wide range of national and international standards for combating illicit trafficking of radiological materials, nuclear, biological and chemicals, there are still regulation gaps that make the illicit trafficking of these materials a reality that must stop. It

> is necessary and urgent to try to fill these regulations gaps in order to ease the fight against this illicit trade for whoever having the competence (2012, p. 93)

A paradigmatic example of the nuclear proliferation of dual-use materials and technology smuggling was the Khan network. This huge and complex network was formed by traffickers and camouflaged companies from different countries. They created a knotty network for the supply and distribution of dual-use components for manufacturing ultra-centrifuges to enrich uranium for Libya, North Korea, and Iran's nuclear programs. The network acted for over 20 years, exporting components from Malaysia through Dubai, evading international laws and governmental controls in three continents. The Khan Network is also a clear example the triangulation as a method of covert acquisition in illicit trafficking.

The International Atomic Energy Agency (IAEA) established an Illicit Trafficking Database (ITDB) with information provided by state members about the illegal acquisition, use, possession, and trafficking of radioactive and nuclear materials. This database reports incidents involving illegal trade and cross-border trafficking, and incidents involving the loss and discovery of uncontrolled radioactive sources. During the period of 1993–2014, a total of 2477 confirmed incidents were reported to the IAEA by participating and some non-participating states. Among these incidents, 16 confirmed incidents involved the unauthorized possession of high enriched uranium (HEU) and high enriched plutonium (HEP), that were acquired in illegal transactions at international borders (ITDB 2016).

The EU has adopted several strategies to enhance CBRN security both at intercommunity and international level: in 2003, the EU Strategy against proliferation of Weapons of Mass Destruction; in 2006, the Instrument for Stability, a CBRN risk mitigation component; in 2008, the New Lines for Action by the European Union in Combating the Proliferation of Weapons of Mass Destruction and their Delivery Systems; and, in 2009, the EU CBRN Action Plan and the review of the EU Dual Use Regulation 428/2009. The main aspects are measures taken to combat intangible transfers of knowledge and know-how, and to intensify efforts to combat proliferation financing, among others.

5 Conclusions

Combatting the illegal use of the internet is a highly challenging task. As we have seen, cyberspace provides an almost unlimited anonymous platform for conducting criminal activities. That is why it is necessary to implement both technological and agreed legal instruments to counteract this threat. A steady reinforcement of international cooperation is also fundamental in the fight against transnational organized crime. This cooperation is essential in the control and surveillance of the dual-use materials trade, especially CBRN dual-use materials, to prevent them from being subject to illegal trafficking and smuggling, as the potential exists for them to be utilized for purposes of proliferation or for terrorist objectives.

References

Bajo M, Bacigalupo S (eds) (2009) Política criminal y blanqueo de capitales. Marcial Pons, Madrid

Battista G (2007) Estrategias de combate de la movilidad patrimonial de la delincuencia organizada transnacional: De la cooperación judicial al reconocimiento recíproco de las órdenes de aseguramiento y decomiso de los productos del delito. In: Herrán M, Santiago JL, González S, Mendieta E (eds) Análisis, Técnicas y Herramientas en el combate a la delincuencia organizada y corrupción con fundamento en la Convención de Palermo. Coyoacán, pp 513–534

Brandaríz JA (2009) Asociaciones y organizaciones criminales. Las disfunciones del art. 515.1º CP y la nueva reforma penal. In: Álvarez García FJ (ed) La adecuación del Derecho penal español al Ordenamiento de la Unión Europea. La Política criminal europea. Tirant lo Blanch, Valencia, pp 725–758

Business Dictionary. http://www.businessdictionary.com

COMJIB (Conference of Ministers of Justice of Ibero-American Countries) (2010) Recommendations on common minimum standards for criminal punishment ofthe organized crime at its 17th Plenary Session, Mexico D.F., 21–22 October 2010

Cuesta Sauquillo MT (2012) Blanqueo de Capitales. In: Magaz Alvarez R (ed) Criminalidad y Globalización: Análisis y Estrategias ante Grupos y Organizaciones al Margen de la Ley. Colección Docencia, IUGM—UNED

EU (2008) New lines for action by the European Union in combating the proliferation of weapons of mass destruction and their delivery systems. www.trade.ec.europa.eu

EU (2009) Regulation (EC) Nº 428/2009. http://eur-lex.europa.eu/LexUriServ/LexUriServ.do?uri=OJ:L:2009:134:0001:0269:en:PDF

EU (2010) CBRN action plan and the review of the EU dual use regulation 428/2009. http://eur-lex.europa.eu/legal-content/EN/ALL/?uri=celex%3A52010IP0467

EU (2011) Directive 2011/93/EU of the European Parliament and of the Council of 13 December 2011 on combating the sexual abuse and sexual exploitation of children and child pornography. http://db.eurocrim.org/db/en/vorgang/204/

EU (2013) Directive 2013/40/EU of the European Parliament and of the council of 12 August 2013 on attacks against information systems and replacing council framework decision 2005/222/JHA. http://itlaw.wikia.com/wiki/Directive_2013/40/EU_of_the_European_Parliament_and_of_the_Council_of_12_August_2013

European Commission (2014) The EU dual use exports control regime. www.trade.ec.europa.etu/doclib/html/152181.htm

European Security Strategy (2016) http://www.eeas.europa.eu/csdp/about-csdp/european-security-strategy/

FATF-GAFI.ORG—Financial Action Task Force (FATF). www.fatf-gafi.org

Fernández Steinko A (2008) Las pistas falsas del crimen organizado. Finanzas paralelas y orden internacional. Catarata, Madrid

Flores CA (2007) Poder, corrupción y delincuencia.Modelo teórico de la articulación de vínculos de contubernio entre funcionarios públicos y el crimen organizado. In: Herrán M, Santiago JL, González S, Mendieta E (eds) Análisis, Técnicas y Herramientas en el combate a la delincuencia organizada y corrupción con fundamento en la Convención de Palermo. Coyoacán, México, pp 141–191

Flores Prada I (2012) Criminalidad informática. Aspectos sustantivos y procesales. Tirant lo Blanch, Valencia

González Mas JL (2012) Instrumentos jurídicos en la lucha contra el crimen organizado. En Criminalidad y Globalización.Análisis y Estrategias ante grupos y organizaciones al margen de la Ley. Ricardo Magaz Alvarez (Coordinador). Colección Docencia. Instituto Universitario General Gutiérrez Mellado (IUGM), pp 143–196

Herrán M, Santiago JL, González S, Mendieta E (2007) Aproximación general a la delincuencia organizada. In: Herrán M, Santiago JL, González S, Mendieta E (eds) Análisis, Técnicas y Herramientas en el combate a la delincuenciaorganizada y corrupción con fundamento en la Convención de Palermo. Coyoacán, pp 21–26

IAEA Incident and Trafficking Database (ITDB) (2016) Incidents of nuclear and other radioactive material out of regulatory control 2016 Fact Sheet. https://www-ns.iaea.org/downloads/security/itdb-fact-sheet.pdf

Informe CGPJ (2008) Al Anteproyecto de Ley Orgánica por la que se modifica la Ley orgánica 10/1995, de 23 de noviembre, del Código Penal. Report on the draft bill of the act of parliament that modified LO 10/1995, of November 23, of the Criminal Code, 2008, pp 49–56, 118, 122

Informe Consejo Fiscal (2008) Sobre el Anteproyecto de Ley Orgánicapor la que se modifica la Ley Orgánica 10/1995, de 23 de noviembre, del Código Penal. Report on the draft bill of the act of parliament that modified LO 10/1995, of November 23, of the Criminal Code, 2008, pp 162–163

Kinzig J (2004) Die rechtliche Bewältigung von Erscheinungs for menor ganisierter Kriminalität. Duncker & Humbold, Berlín

Manjón Cabeza A (2010) Receptación y blanqueo de capitales (arts. 301 y 302). In: Álvarez García FJ, González Cussac JL (eds) Comentarios a la Reforma Penal de 2010. Tirant lo Blanch, Valencia, pp 339–346

Mapelli B (2001) Estudios sobre delincuencia organizada. Instituto Andaluz Interuniversitario de Criminología, Sevilla

Ministerio de Comercio. Gobierno de España (2008) Royal Decree 2061/2008 of 12 December approving the control Regulation on foreign trade in defence material, other material and dual-use items and technologies. http://www.comercio.gob.es/en/comercio-exterior/informacion-sectorial/material-de-defensa-y-de-doble-uso/PDF/concepts/20090130RD20612008EN1.pdf

Miró F (2012) El cibercrimen. Fenomenología y criminología de la delincuencia en el ciberespacio. Marcial Pons, Madrid

Moore M (1986) Organized crime as business enterprise. Major Issues in organized crime control. Symposium conducted at the meeting of the US Department of Justice, Washington, DC

Morán Rubio JL (2012) Tráficos ilícitos: Especial referencia a materiales de doble uso. In: Magaz Alvarez R (ed) Criminalidad y Globalización: Análisis y Estrategias ante Grupos y Organizaciones al Margen de la Ley. Colección Docencia, IUGM—UNED

Otero MP (2011) Possible criminal justice solutions to organized crime: drug trafficking and terrorism. Open Criminol J 4:109–119

Otero MP (2013) The phenomenon of terrorism as organized crime: reasons for specific punishment and punitive responses. An Spanish law perspective. In: Walters TK, Monaghan R, Martín Ramírez J (eds) Radicalization, terrorism, and conflict. Scholars Publishing, Cambridge, pp 104–129

Rivera J. Blue-collar definition. http://www.legalmatch.com/law-library/article/what-is-blue-collar-crime.html#sthash.Xi37UC5.dpuf

Spanish Parliament (1995) Ley Orgánica 12/1995 de Represión del Contrabando. https://www.boe.es/boe/dias/1995/12/13/pdfs/A35701-35705.pdf

Spanish Parliament (2007) Ley 53/2007. https://www.boe.es/buscar/pdf/2007/BOE-A-2007-22437-consolidado

Spanish Parliament (2011) Ley Orgánica 6/2011. https://www.boe.es/diario_boe/txt.php?id=BOE-A-2011-11264

Spanish Parliament (2015) Ley Orgánica 2/2015 of the modification of the Criminal Code (BoletínOficial del Estado, nº 77, 31-marzo-2015), p 27177

Toval L (2012) Fenomenología del Crimen OrganizadoTransnacional. In: Magaz Alvarez R (ed) Criminalidad y Globalización: Análisis y Estrategias ante Grupos y Organizaciones al Margen de la Ley. Colección Docencia. IUGM—UNED

United Nations (2005) Resolution 1540/2005 of the United Nations. http://www.un.org/en/sc/1540/

United Nations (2014) U.N. Security Council Resolution 2178, September 24, 2014
United Nations Office on Drugs and Crime (2009) Transnational organized crime: the globalized illegal economy. https://www.unodc.org/toc/en/crimes/organized-crime.html
Zúñiga L (2002) Redes internacionales y criminalidad: a propósito del modelo de "participación en organización criminal". In: Zúñiga L, Méndez C, Diego MR (eds) El Derecho penal ante la globalización. Colex, Madrid

Author Biographies

Natividad Carpintero-Santamaría is Full Professor at the Department of Energy Engineering of the Polytechnic University of Madrid, General Secretary of the Institute of Nuclear Fusion and Member of the Presidium of the European Academy of Sciences. She holds a Diploma in High Studies of Defence and a Diploma as University Expert in Transnational Organized Crime and Security. She has been member of the Consulting Board of the International Working Group of the G8 Global Partnership. She is a collaborator in CBRN threats research at the Spanish Centre for National Defence Studies and other institutions. She has lectured in different countries: Armenia, Australia, EU, Latin America, Middle East, US and Russian Federation and published several papers on asymmetric threats, WMD terrorism, illicit trafficking of radioactive materials and energy security. She has been granted the Cross of Aeronautical Merit (white distinctive) and the Cross of Military Merit (white distinctive).

María Pilar Otero is Full Professor at the Department of Criminal Law of the Carlos III University of Madrid and currently Vice-Dean at the School of Law of the same University. She has worked on criminal protection of different types of secrets and professional confidentiality, as well as on criminal protection of cultural property. She has also tackled subjects of corruption. Another of her fields of research is gender criminal law, studying sexual harassment and domestic violence. She is currently working on different types of sentences in the field of prison law, such as home confinement or electronic methods of control and their efficacy regarding control of aggression and of social rehabilitation of the offender. Additionally, she has been named Expert Technical Consultant in Criminal Law by The European Union, supporting legal modernization in Paraguay and Dominican Republic.

Some Criminal Aspects of Cybersecurity

J. Martín Ramírez

Abstract These pages are meant to be a modest step in order to help the prevention and the combat of cybercrime, raising awareness that the birth of the ubiquitous "cyber spectrum" brings with it a parallel presence of plenty of 'new' crimes. The cybersecurity ecosystem, besides obvious benefits, also presents new potential specific cybercriminal threats, criminal opportunities, and vulnerabilities, as well as physical and psychological harm to individuals. After describing the cybercrime phenomenon and enumerating the spectrum of it; the importance of cyber-security awareness through public education stressing emerging threats and risks is highlighted in the hope that individual internet users will take basic security precautions. CYBERCRIME IS SILENT VIOLENCE, THEREFORE ME MUST PREVENT CYBERCRIMES!!!

Keywords Cyberspace · Cybersecurity · Cybercrime

1 Cyberspace and Cybersecurity

The presence of Information and Communication Technologies (ICTs) in the world is increasing in geometrical proportions. The few employers who were using the old IBMs half a century ago have multiplied by millions to become now a world

The main points of this chapter were elaborated during the author's leave at Stanford University, as a Visiting Fellow of the Hoover Institution on War, Revolution and Peace.

Submitted: 30.7.2016; Accepted: 1.9.16.

J.M. Ramírez (✉)
Hoover Institution, Stanford University, Stanford, USA
e-mail: mramirez@ucm.es
URL: http://www.jmartinRamirez.org

J.M. Ramírez
Nebrija-Santander Chair on Risk and Conflict Management, Nebrija University, Madrid, Spain

J.M. Ramírez and L.A. García-Segura (eds.), *Cyberspace*, Advanced Sciences and Technologies for Security Applications, DOI 10.1007/978-3-319-54975-0_8

characterized by webs and clouds, with hundreds of millions of Internet users. In 2013, at least 2.3 billion people—equivalent to more than one third of the world's total population—had access to the Internet, and the number of networked devices, what experts already denominate the "Internet of Things" (IoT), are estimated to outnumber people by six to one, transforming current conceptions of the Internet. The United Nations Office on Drugs and Crime (UNODC) (2013) estimated that mobile broadband subscriptions would approach 70% of the world's total population by 2017. Another indication of the growing importance of IoT is that the more than 6.4 billion connected objects in use in 2016 are expected to explode by 2020, where the market will include 20.8 billion things, accounting for over two-thirds of the worldwide IoT market, according to a Gartner report (2015).

This artificial sphere created by informatics, known as cyberspace, is a virtual reality, a digital world characterized by telecommunication webs and a series of devices and information systems interconnected through them. It overlaps the other four classical domains, namely land, sea, air, and outer-space, becoming the 'fifth domain'.

Of course, this hyper-connected world represents an immense source of knowledge and wellbeing for humanity. But, at the same time, these technological and operational improvements mean that risks and threats have also evolved, taking advantage of all the vulnerabilities of the webs and their information systems. Crooks and spooks are finding plenty of chinks in the digital armor. The lonely, romantic hackers of the 1990s, whose only aim seemed to show their technical capabilities and to get public relevance, changed with the turning of the century into organized crime (Otero 2013). In the future hyper-connected society, it is hard to imagine any crime that will not involve electronic evidence linked with Internet protocol (IP) connectivity. For instance, the IoT, interconnecting the most quotidian objects via computers and networks, installing internet-connected sensors, is doubtless a very promising technological step towards a more efficient world, benefiting especially—I would add—those giants intermediaries who apply these new artificial-intelligence capabilities for collecting and analyzing huge chunks of data. But the IoT also offers more open gates towards vulnerability and insecurity. New technologies thus create new criminal opportunities. Therefore, we are in the presence of an emerging priority challenge (García-Segura and Ramírez 2016).

In this context, cybersecurity is a very important challenge. Risks in cyberspace are some of the main priorities of individuals, corporations, or governments because the wrong use of these critical infrastructures may lead to authentic threats to our economy, our productivity, and our security. Consequently, it is very necessary to take steps to enhance the security of ICTs in front of eventual crimes in the cyberspace (Ramírez and Alfaro 2013). Cybercrime is advancing in the focus of the public due to increased media reporting of cybercrime cases, cybersecurity issues and other cyber-related news, aware of its dangers and even of its potentially catastrophic consequences (Lucas 2015).

2 Describing and Categorizing Cybercrime

Numerous academic works have attempted to define cybercrime (International Telecommunication Union 2011; Pocar 2004; Wall 2007). In a wider sense, cybercrime ('computer crime') encompasses any illegal behavior committed by means of electronic operations that targets the security of computer systems or networks and the data processed by them. Additionally, cybercrime also includes traditional crimes conducted through the Internet. For example; hate crimes, telemarketing and Internet fraud, identity theft, and credit card account thefts are considered to be cybercrimes when the illegal activities are committed through the use of a computer and the Internet (Webopedia 2014).

At the core of the cybercrime concept lays an attack against the confidentiality, integrity, and availability of computer data or systems belonging to governments, corporations, and citizens. These attacks do not take place on a physical body but on the personal or corporate virtual body, which is the set of informational attributes that define people and institutions on the Internet. In other words, in the digital age, our virtual identities are essential elements of everyday life: we are a bundle of numbers and identifiers in multiple computer databases owned by governments and corporations.

What distinguishes cybercrime from the traditional criminal activity? Criminals do not need a computer to commit fraud, traffic in child pornography and intellectual property, steal an identity, or violate someone's privacy. All those activities had existed before the "cyber" prefix became ubiquitous. Obviously, one difference is the use of the digital computer, but technology alone is insufficient for any distinction that might exist between different realms of criminal activity. Cybercrime, especially involving the Internet, represents an extension of existing criminal behavior alongside some novel illegal activities. Another characteristic is that the Internet provides relative anonymity to criminals.

Cybercrime ranges across a spectrum of activities. At one end are crimes that involve fundamental breaches of personal or corporate privacy, such as assaults on the integrity of information held in digital depositories and the use of illegally obtained digital information to blackmail a firm or individual. Also at this end of the spectrum is the growing crime of identity theft. Midway along the spectrum, lyes the transaction-based crimes such as fraud, trafficking in child pornography, digital piracy, money laundering, and counterfeiting. Besides these specific crimes with specific victims, other crimes involve individuals within corporations or government bureaucracies deliberately altering data for either profit or political objectives. A clear, present example may be the known WikiLeaks case, and more specifically the revelations of Edward Snowden, leaking details of American surveillance operations to the media, and the most recent US Democratic Party e-mail scandal.[1]

---

[1]In March 2015, it became publicly known that Hillary Clinton, during her tenure as United States Secretary of State, had exclusively used her family's private email server rather than official State Department accounts maintained on federal servers, for thousands of emails that would later be

At the other end of the spectrum are those crimes that involve attempts to disrupt the actual workings of the Internet. These range from spam, hacking, and denial of service attacks against specific sites to acts of cyberterrorism—that is, the use of the Internet to cause public disturbances and even death (Encyclopedia Britannica 2014).

At the end of the last century, Grabosky and Smith (1998) listed nine types of cybercrime: 1. theft of telecommunications services; 2. communications in furtherance of criminal conspiracies; 3. telecommunications piracy; 4. dissemination of offensive materials; 5. electronic money laundering and tax evasion; 6. electronic vandalism, terrorism and extortion; 7. sales and investment fraud; 8. illegal interception of telecommunications; and 9. electronic funds transfer fraud.

A couple of years later, the Council of Europe promulgated a Convention on Cybercrime (2001), where the different kinds of cybercrime were organized into three broad categories: 1. Offences against the confidentiality, integrity and availability of computer data and systems: illegal access and interception, data and system interference, and misuse of devices; 2. Computer-related offences: forgery and fraud; and 3. Content-related offences: child pornography, infringements of copyright and related rights, search and seizure of stored computer data, mutual assistance, spontaneous information, confidentiality and limitation on use.

In a more recent comprehensive study on the topic, the United Nations Office on Drugs and Crime (UNODC 2013) has categorized a none-exhaustive list of 14 offenses that may constitute cybercrime. Most of the crimes coincide with the ones included in the above mentioned Council of Europe Convention on Cybercrime, plus a few new ones (they are highlighted in italics):

1. A limited number of acts against the confidentiality, integrity, and availability of computer data or systems represent the core of cybercrime
   - Illegal access to a computer system
   - Illegal access, interception or acquisition of computer data
   - Illegal interference with a computer system or computer data
   - Production, distribution or possession of computer misuse tools
   - *Breach of privacy or data protection measures.*
2. Computer-related acts for personal or financial gain or harm
   - Computer-related fraud or forgery
   - *Computer-related identity offences*
   - *Computer-related copyright or trademark offences*

(Footnote 1 continued)

marked classified official communications. On July 5, 2016 the FBI stated that at least Clinton was "extremely careless" in handling her email system (see https://en.wikipedia.org/wiki/Hillary_Clinton_email_controversy).

- *Sending or controlling the sending of spam*
- *Computer-related acts causing personal harm*
- *Computer-related solicitation or 'grooming' of children.*

3. Computer content-related acts

- Computer-related acts involving *hate speech*
- Computer-related production, distribution or possession of child pornography
- Computer-related acts in *support of terrorism offences.*

Summarizing, the following main categories of cybercriminal threats may be envisaged: intrusion for monetary or other benefit, interception for espionage, manipulation of information or networks, data destruction, misuse of processing power, counterfeit items, and evasion tools and techniques.

This picture adds up to one of a complex balance of the current cybercriminal threat landscape, which has been analyzed in detail by the International Cyber Security Protection Alliance (ICSPA), member organization Trend Micro and the European Cybercrime Centre (EC3) at Europol under the following topics: cloud/virtualization, consumerization/bring your own device (BYOD), crime as a service, cyber weapons, data-stealing Trojans, embedded hardware, hacktivism, high profile data loss, industrial control systems (SCADA), legislation working against security, malware outside the operating system mobile, new threat actors, new ways to hide, online financial service attacks, rogue certificates, social engineering, social networking/media, spam goes legitimate, secure sockets layer (SSL) and transport layer security (TLS) attacks, targeted attacks, and web exploits (for a detained analysis of this landscape see: ICSPA 2012). This chapter enumerates them not only because they serve as a general set of current indicators for the evolution of cybercrime, but also because they enable the identification of a number of horizontal trends to be developed out in the future.

3 A Challenging New Future: Growing Risks

How can we prepare ourselves for the present and future challenges of cybercrime? From the specific perspectives of each of us—an ordinary Internet user, a manufacturer, a communications service provider, or a government-, we all are required to be aware of the growing new risks, such as some increasingly blurred distinctions in relation to traditional crimes, particular crime prevention challenges of plausible and possible future impacts of cybercrime, and finally some specific legal concerns.

3.1 Some Blurred Distinctions

There is an increasingly blurred distinction between legitimate and illegal activity, between misuse and legitimate use. How will we achieve consistency? What data will the authorities be able to access and use for the purposes of preventing and disrupting criminal activity?

Evolved threats to critical infrastructure and human implants will increasingly blur the distinction between cyber and physical attack, resulting in offline destruction and physical & psychological harm to individuals.

Although human rights standards have to be applied equally in the context of computer data and electronic communications by means of a wide range of safeguards for the protection of privacy; the increasing impact of cloud computing services on security considerations introduces a high degree of intrusiveness and uncertainty for users concerning the privacy regime that will apply to their data[2]—for instance, users are not always informed exactly 'where' their data is held, and the circumstances under which privacy may legitimately be infringed for the purposes of law enforcement investigations or security surveillance.

As the world moves towards universal Internet access, conceptions of cybercrime may need to operate on a number of levels: specific and detailed in the case of the definition of certain individual cybercrime acts, but sufficiently broad to ensure that investigative powers and international cooperation mechanisms can be applied, with effective safeguards, to the continued migration of offline crime to online variants.

3.2 Cybercrime Prevention

The Internet has created the greatest treasure trove of personal data ever assembled. Such data may be legitimately used by governments, NGOs, and companies, and even be invaluable for companies that want to target their marketing more closely. But this raises serious questions, such as: who should be responsible for the data's security? Or should there be limits on how they can be used? Because neither private companies are always trustworthy, nor unscrupulous or careless governments can be ruled out (Economist 2014). We should never forget that this collection of oodles of information about everyone may erode individual privacy which has to be protected (Angwin 2014; Cayón and Cortijo 2016).

The flexibility of Internet, with new sites springing up all the time, makes it very difficult to monitor. This fact presents particular crime prevention challenges, such as the increasing ubiquity and affordability of online devices leading to large

[2]This can occur, for example, where cloud computing providers store data in multiple copies in data centers in different countries, and make use of dynamic data management between these data centers.

numbers of potential victims; the comparative willingness of persons to assume 'risky' online behavior; the possibility for anonymity and obfuscation techniques on the part of perpetrators; the transnational or global nature of many cybercrime acts; and the fast pace of criminal innovation (UNODC 2013).

Crime prevention plans seek to reduce the risk of crimes occurring, and their potentially harmful effects. A good strategy for this emphasizes a range of interactions between governments, the private sector, and academia, establishing public-private partnerships for the exchange of good practice for the prevention and combat of cybercrime.

At the government level, the OECD has prepared a "Guidelines for the Security of Information Systems and Networks" (OECD 2002), reflected in a General Assembly Resolution by the United Nations concerning the creation of a Global Culture of Cybersecurity (United Nations 2003).

The private sector shows the following trends: (a) a range of cybersecurity awareness and actions—while threat awareness is growing, this does not immediately lead to behavior change though; (b) use of cybersecurity technology, to prevent cybercrime, such as firewalls, digital evidence preservation, and restrictions on specific IP address connections; and (c) take proactive steps to counter cybercrime acts, investigating and shutting down cyberattacks that threaten consumer trust in their systems, such as spam e-mails and other unsolicited communications (Sorkin 2001), malicious software and botnets, or even 'hacking back' against attackers (Higgins 2012), or mimicking eventual cyber-invaders and testing the vulnerabilities of clients' systems—a practice known as "penetration testing", or conducting regular "pentests".

Academic institutions represent an important partner in cybercrime prevention through these ensuing actions: (a) knowledge development and sharing; (b) legislation and policy development; (c) the development of technology and technical standards; (d) the delivery of technical assistance and (e) cooperation with law enforcement authorities.

Just to add that, for a timely prevention or response to cybercrime, faster informal initiatives, and cooperation among institutions are conveniently needed due to the volatile nature of electronic evidence. For instance, where there is an imminent threat of harm, urgent access to cloud data can be required.

3.3 Some Specific Legal Concerns

In order to be capable to investigate and combat such 'novel' cybercrime threats, more creative and flexible responses to criminality have to be developed, enhancing the capacities of law enforcement and the criminal justice system, and requiring technical assistance activities in accordance with the specificities of cybercrime.

The growth of the Internet, making it much easier for both companies and the government to gather large amounts of information about individuals, leads the state into the temptation of taking advantage of this corporate knowledge, largely for policing and anti-terrorism reasons, and also to feel the need to regulate how companies store and use personal data.

There is also a growing legal fragmentation of the instruments aimed at countering cybercrime, in terms of cybercrime acts, jurisdictional bases, and mechanisms of cooperation (Koops 2010). Consequently, new concepts and objects, such as intangible 'computer data,' not addressed by 'traditional' law, establishing specialized offences for a set of cybercrime acts, as well as the promulgation of multilateral legal instruments, inherently designed to play a role in harmonization of cybercrime laws for, inter alia, the elimination of criminal safe havens, and global evidence collection, are essential today (Calderoni 2010; Otero 2013; Segura-Serrano 2006). Subsequently, criminal law, which historically regulates interactions between human beings, will inevitably have to be enhanced to unmanned vehicles, robotic devices and automation (Tejada de la Fuente 2016).

One key external standard that offers guidance in this area is the contribution of international human rights law in assisting states to achieve an acceptable balance between crime prevention and control, and the protection of individual liberties. Some prohibitions on freedom of expression, including incitement to genocide, hatred constituting incitement to discrimination, hostility or violence, incitement to terrorism, and propaganda for war, are therefore required for states that are party to relevant international human rights instruments. For other forms of expression, the 'margin of acceptable expression' is in line with their own standards, cultures, and legal traditions. International human rights law thus helps to delineate acceptable expressions acting both as a sword and a shield, requiring criminalization of extreme forms of expression, while protecting other forms (Ramírez 2014; UNODC 2013).

4 Conclusion

In conclusion, this hyper connected world with truly immersive technologies, such as wirelessly medical implants, defibrillators, pacemakers, insulin pumps, remote presence, virtual reality and sensor technologies already coming on to the mainstream market, represents an immense source of benefits for humanity. But, at the same time, it is also reasonable to speculate that its level of interaction of human cognitive processes presents new growing risks—such as potential specific cyber-criminal threats, criminal opportunities and vulnerabilities harms, which may bring new harms (especially psychological) as well. We have to confront these challenges with exquisite prudence, not only because one should always have in mind the dilemma of security versus freedom, but also to avoid falling into unnecessary paranoia; if user costs are higher than direct user benefits, than individuals have a strong incentive to ignore security measures. The use of any tool implies some level of risk, but it would be better than to do nothing (Ramírez and Alfaro 2013).

The considerations presented along these pages want to highlight the importance of cyber-security awareness through public education stressing risks (United Nations Guidelines for the Prevention of Crime 2002; Galexia 2011); hopefully help individual Internet users in taking basic security precautions,[3] and to educate them on the emerging threats, as well as be aware of those targeted as specific audiences, such as children.

Acknowledgements Thanks are due to the Hoover Institution on War, Revolution and Peace, at Stanford University, for the facilities given to the author in order to write this paper during his stay, as a Visiting Fellow. Part of it was used during an unpublished lecture on November 3rd, 2013 at Istanbul (Turkey), on the occasion of the 60th Pugwash Conference on Science and World Affairs.

Conflict of Interest
The author confirms that this article content has no conflicts of interest.

References

Angwin J (2014) Dragnet nation: a quest for privacy, security and freedom in a world of relentless surveillance. Times Books

Calderoni F (2010) The European legal framework on cybercrime: striving for an effective implementation. Crime Law Soc Change 54(5):339–357

Cayón J, Cortijo B (2016) The dilema security vs. freedom. In: 41st CICA on Mapping the Cyberspace, Nebrija University, Madrid, 1–3 June 2016

Council of Europe (2001) Treaty No. 185 Convention on Cybercrime, Budapest, 23.XI.2001. https://www.coe.int/en/web/conventions/full-list/-/conventions/treaty/185

Economist (2014) Governments' relationship with the tech sector is hideously complicated. Econ, 22 Feb 2014

Encyclopedia Britannica (2014) Retrieved from: http://global.britannica.com. (24 Feb 2014)

Galexia (2011) An overview of international cyber-security awareness raising and educational initiatives. Australian Communications and Media Authority

García-Segura LA, Ramírez JM (eds) (2016) Mapping the cyberspace. An emerging priority challenge. Universidad Antonio de Nebrija, Madrid

Gartner Inc. (2015) 6.4 billion connected "things" will be in use in 2016, up 30 percent from 2015, Gartner report November 10, 2015STAMFORD, Conn., http://www.gartner.com/newsroom/id/3165317

Grabosky P, Smith R (1998) Crime in the digital age. Federation Press, Sydney

Herley C (2009) So long, and no thanks for the externalities: the rational rejection of security advice by users. New Security Paradigms Workshop, Oxford

Higgins KJ (2012) Turning tables: ID'ing the hacker behind the keyboard. Dark Read, 2 Oct 2012

International Cyber Security Protection Alliance (ICSPA) (2012) Project 2020. Scenarios for the future of cybercrime—white paper for decision makers. European Cybercrime Centre (EC3), Europol

[3]"If users spent even a minute a day reading URLs to avoid phishing, the cost (in terms of user time) would be two orders of magnitude greater than all phishing losses (Herley 2009)".

International Telecommunication Union (2011) Understanding cybercrime: a guide for developing countries. Explanatory Report to the Council of Europe Cybercrime Convention, ETS No. 185
Koops B-J (2010) Cybercrime legislation. Electron J Comp Law 14(3)
Lucas E (2015) Cyberphobia: identity, trust, security and the Internet. Bloomsbury
OECD (2002) Recommendation of the council concerning guidelines for the security of information systems and networks—towards a culture of security, OECD, 25 July 2002
Otero P (2013) Terrorism and organized crime in the Spanish criminal law: reasons for specific punishment and punitive responses. In: Walters TK, Monaghan R, Ramírez JM (eds) Radicalization, terrorism, and conflict. Cambridge Scholars Publishing, Newcastle, pp 104–129
Pocar F (2004) New challenges for international rules against cyber-crime. Eur J Crim Policy Res 10(1):27–37
Ramírez JM (2014) Moving toward peace. In: Ramírez JM, Morrison C, Kendall AJ (eds) Conflict, violence, terrorism, and their prevention. Cambridge Scholars Publishing, Newcastle, pp 191–206
Ramírez JM, Alfaro PJ (2013) Cyberspace and cyber attacks. In: 60th Pugwash conference on science and world affairs on dialogue, disarmament, & regional and global security, Istanbul, Turkey, 1–5 November 2013
Segura-Serrano A (2006) Internet regulation and the role of international law. In: Von Bogdandy A, Wolfrum R (eds) Max Planck Yearbook of United Nations Law 10:191–272
Sorkin DE (2001) Technical and legal approaches to unsolicited electronic mail. Univ San Francisco Law Rev 35(2):359–360
Tejada de la Fuente E (2016) Legislative reform on the fight against cybernetic crimes. In: 41st CICA on Mapping the Cyberspace, Nebrija University, Madrid, 1–3 June 2016
United Nations (2002) Guidelines for the prevention of crime. Economic and Security Council resolution 2002/13
United Nations (2003) General assembly resolution 57/239, 31 Jan 2003
UNODC (United Nations Office on Drugs and Crime) (2013) Comprehensive study on cybercrime. United Nations, New York. http://www.unodc.org/documents/organized-crime/UNODC_CCPCJ_EG.4_2013/CYBERCRIME_STUDY_210213.pdf
Wall DS (2007) Cybercrime: the transformation of crime in the information age. Polity Press, Cambridge
Webopedia (2014) Cybercrime definition. Retrieved from: http://www.webopedia.com/TERM/C/cyber_crime.html (21 Feb 2014)

Author Biography

J. Martín Ramírez Present Chair of the Nebrija-Santander Chair on Risk and Conflict Management at Nebrija University, previously he was head of the Complutense Research Group on Sociopsychobiology of Aggression, as well as of the Departments of Psychobiology at Seville and Complutense Universities, and of the Rector Office at the Autonomous University of Madrid. He studied Medicine, Humanities, and Law, obtaining a Ph.D. degree in Medicine and Surgery (Neurosciences) and in Philosophy (Education), as well as three Diplomas and a Master in National Defense at the Spanish CESEDEN. Research Fellow of the International Security Program of the Harvard University and Visiting Fellow of the Hoover Institution on War, Revolution and Peace, Stanford University, among other institutions. Several times official of the Boarding Conseil of the International Society for Research on Aggression (ISRA), he is on the Advisory Board of the Society for Terrorism Research and of the Professors World Peace

Academy, and is author of more than 450 scientific publications, among books, articles, and scientific communications. He is also a Fellow of the World Academy of Art and Science, a member of the New York Academy of Sciences, and Chair of CICA International and of the Spanish Pugwash Movement (Peace Nobel Price 1995).

The Situation and Evolution of the Managed Services of Cybersecurity, Towards 3.0 and Beyond

Juan Miguel Velasco

Abstract The evolution of the digital world and the entry of the Cloud, the Internet of Things (IoT), the mobile applications and the Big Data among others, demand upon the providers of programs and services of Security an urgent evolution. It is the new Cybersecurity 3.0, services without digital or geographical boundaries, with no perimeter to secure and with reach in all the technological and analog areas, an evolution that many users (CSO and CISOs) have not yet seen. From the evolution of services 1.0 to Cybersecurity 3.0, this article covers the types of services and the characteristics they must have in order to anticipate and provide effective solutions in the new environments.

Keywords Managed security services · Cloud security · Perimeter security · Security outsourcing

1 Security Services Generation 1.0: From Mediocrity to Taximeter

After almost 25 years in Information technology (IT) and the last 17 of them in Security in Large Corporations, one gets to the idea of how is the panorama of managed services. Today we "enjoy" some sort of "managed security" providers, where unfortunately most of them are really body-shopping companies, as we say here, quite badly indeed, since it is a word with which one should be careful and never use in the USA or UK, because it has other connotations more related to

Submitted: 26.11.16; Accepted: 30.12.16.

J.M. Velasco (✉)
Aiuken Solutions, Madrid, Spain
e-mail: jmvelasco@aiuken.com

J.M. Ramírez and L.A. García-Segura (eds.), *Cyberspace*, Advanced Sciences and Technologies for Security Applications, DOI 10.1007/978-3-319-54975-0_9

human trafficking than to IT, although I fear that it is often a true reflection of the reality of the service it describes.

In short, most large suppliers, with big names and enormous capacities on paper, combine disparate technicians that are not very well trained to go to the customer to "manage" security, and we often do not know if it is technically capable or they are like Gary Cooper in the film "High Noon", where he had to face alone with all the "bad guys" to manage all the systems and platforms with no knowledge at all.

That is not managing security, but it is aim to profit from the administration of security systems with elegance and with lots of luck not to have incidents, and they keep their finger crossed so there are no attacks or that nobody breaks anything Intentionally or not.

These outsourcing services are not Managed Security, and although their price may be competitive, they only offer Operational hands to the Client to manage without a security government, nor to worry about the Cybersecurity strategy, nor policies, nor Security Operations Procedure (SOP) in production and the don't use the Plan Act Do Check method (PADC) at all. They are services that have all the big suppliers of IT, Integrators, Telecommunications providers and consulting or audit firms, but are those really services? In fact, they are dedicated projects, custom made projects, very well armed, and in fact their effectiveness and success in the people who dedicate themselves to it, in many cases those professionals are collected or subcontracted by indication of their end user himself.

The reality is that a service is something common, replicable, scalable and oriented or customized to the clients, such as its definition by the Spanish Royal Academy of Language dictionary indicates. Figure 1 shows the common security service provider platform elements that should be counted as a minimum base. These elements are essential to provide as shared knowledge and a management platform. Without them, it is impossible to provide a security management, not an administration of security systems.

The difference is very relevant, since a good provider of Managed Security Services, must meet at least a minimum to be so named:

1. It must support ALL security technologies that the client has.
2. It must respect the freedom of the customer to choose security technologies and decide according to his own criteria (budget, reliability, closeness, etc.).
3. It must be proactive in management, critical and analytical in administration (it is not worth applying tons of security rules without control or reasoning).
4. It must know the environment and have information on vulnerabilities and the context of technologies managed in isolation and as a whole.
5. It must have its own management, monitoring and administration tools.

Suppliers who mix body-shopping services with IT management will call them Suppliers 1.0. These 1.0 providers offer their basic package of services (see Fig. 2), which are always directly related to IT management. We have the management and administration of any security element: Managed FW (Firewall), Managed IPS (Instructions Prevention System), Managed proxy, etc. They really do not lie when they say that they are Managed Security providers, because what they do is manage,

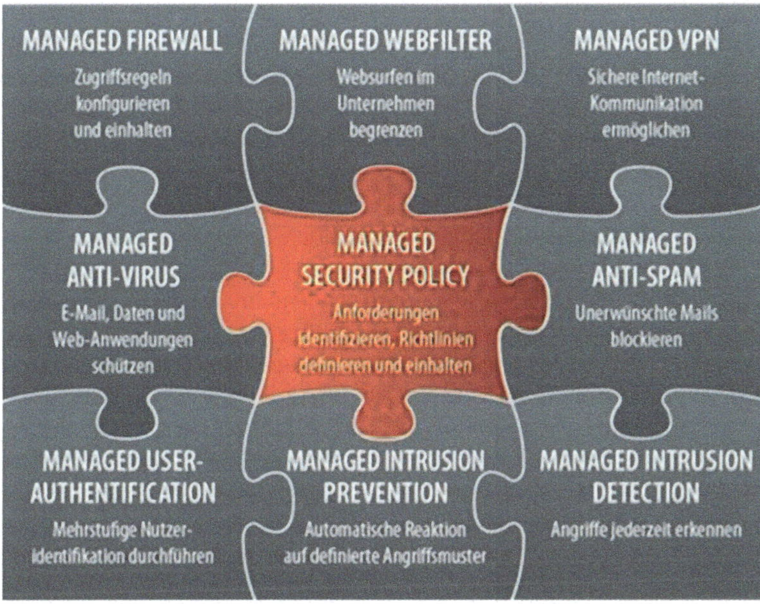

Fig. 1 This picture shows the traditional standard MSSP Offering 1.0

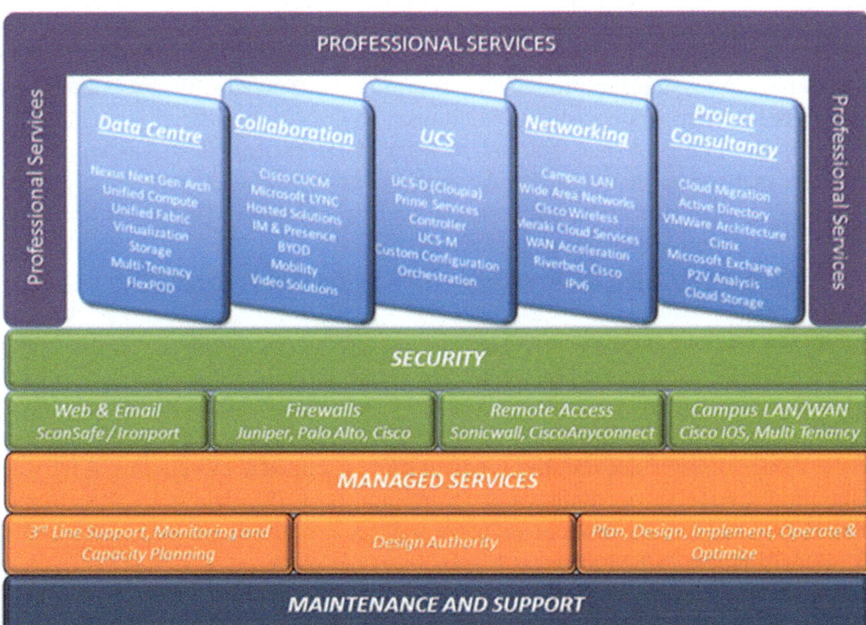

Fig. 2 Example of Advanced portfolio for 2.0 Managed Security Provider

that in the jargon of SOC 1.0 is that things work, do not give noise and do not fall, yes, if there are 20 rules in the firewall with ANY *. * That is something of the client or others, because they manage, which are still level 1 operator services illustrated, something like a robot administrator. What they bring is low cost for the client, in OPEX mode (expense) that normally helps to improve the results account. When there is an incident, which may be inevitable, they are put into taxi mode and the overtime and fire extinguishing actions are triggered, with a significant price impact for the affected client, since they are often incidents as they say "outside Contracted baseline".

This has taken us from the mediocrity of the beginnings, 10 or 12 years ago, in which the security administration was separated from the administration of systems by only a few initial policies and procedures, to the taximeter model in which the security providers charge by people, devices and years, just like a taxi driver that takes you from Downtown to City limits doing nothing more than driving, without worrying about the route, elements of the route, previous driving data, traffic or incidents, as well as passenger knowledge. In addition, many integrators are more concerned with selling the shock absorbers and wheels than the taxi transport service itself; that is, they focus on one or another technology without analyzing the needs of customers, and the necessary multi-technology and multi-layer essential in the development of a good security strategy, based on redundancy and segregation of risks and functions.

These services are characterized by the absolute disregard for innovation or the improvement of new technologies, they are Security Operation Centers (SOCs), with zero level of innovation or improvement, no platform or technology renewals, no test laboratories and pre-production, to promote improvements and suggestions in the cybersecurity strategy of the customers.

That is the future of that taxi and of the Cybersecurity 1.0: to advance to be conscious that service is not "to guide a car by the highway" and to react to what happens, but to plan, to prevent, to know and to anticipate. That's the Managed Security that customers need today and tomorrow.

It is not surprising that many customers are still afraid to get into the taxi (they are more scared than in their own car); that is, they do not want to outsource or support part of their services in external, since they do not see that the "taxi", and they do not follow the driving rules they want. It is important therefore to guide the customer, but much more to listen and give him the service to make him feel comfortable with it. Besides of that, prices often do not invite to use services, but to use the vehicle itself, since security involves three parameters closely linked the rule of the three Cs (CCC): confidentiality, trust[1] and cost. But definitely, if we move forward Taxi to a service of more value we will see the advantages of Cybersecurity 2.0, we will have reached the next level 2.0, to the plane.

---

[1] *Confianza*, in Spanish: this explains the three Cs.

2 Security Services Generation 2.0: From Taxi to Airplane

The options of security providers in our market are not only mediocre options of "body-shopping" model, since there is a group of serious and pioneering providers, professionals in our market, that focus on security at a global level, as a system of Intelligence, anticipation and knowledge. We will call these providers 2.0, Advanced offer Cybersecurity 2.0 services. Let's see how they do it, and because those services are light years away from their fellow Cyber 1.0.

Same as the taxi that only knows the destination and starts and drives, Cybersecurity 2.0. It is more like the plane, which knows the destination but it plans the route, the climatological elements (wind, pressure, clouds, rain, etc.), the elements of the service like the passenger number, its criticality, the merchandise, the weight of the aircraft, and the combination of all, and also on the road follows this continuous analysis of the context to act and anticipate adverse elements (climatology, route, etc.).

The providers of Security Services Cybersecurity 2.0 are characterized by the differentiation in two key elements: the strategy and the intelligence.

Cybersecurity 2.0 Strategy, the strategy in a set of elements of anticipation relative to the security, that must complement the services of the catalog of security of the supplier. A good strategy should include at least the following elements to reinforce Security services:

1. Continuous knowledge of the national and global manufacturers market
2. Management of shared configurations and policies for all clients (no islands per client), and in the wider context of the manufacturers
3. Customer platforms innovation laboratory (real pre-production)
4. Stress laboratory tests and cyber-exercises of each client (democenter of tests per client)
5. Network Knowledge Exchange (FIRST, CERT Network, APWG, iSMS, EuroCloud, ISACA, etc.) and security forums Rooted, BlackHat, Black Knives, DefCon (at least you must speak English and Russian…)
6. Third party (security) certifications: 27.001, LEET Security, 25999, etc.
7. Globality (SOC Multi-territory connected).

We will briefly develop each point so as not to lengthen it:

Point 1, global relationships with local manufacturers, tests, demos, innovation analysis, congresses and models are essential. In IT security and physical security all knowledge of innovation is always little. Being conservative is good in execution and operation but not in strategy and organizations are too busy to innovate, for that we are paid to security vendors to always go ahead (Fig. 3).

Points 2 and 3 are linked to 1 so that the client always has a testing and innovation environment, and can see in the most advanced case the impact of those technologies and policies in their environment as an advance in time, as well as the air route predictive systems that give the best route based on the estimated climatic conditions to be more efficient without fear of innovating in a controlled way.

Fig. 3 No need for innovation, really?

Point 4: it's mandatory that an advanced MSSP 2.0 or higher explore new techniques, technologies and practices, so it's highly recommended that this MSSP build and explore in his own laboratory. It's recommended that they deploy at least all vendors and technologies that runs in production for their customers.

Point 5 is a basic pillar, to evaluate a local service, since the global connection is FUNDAMENTAL, threats, Z-Days, rootkits, botnets etc. They are global and the dark-side community is cooperative, so it is necessary to be connected to national and international knowledge networks to share and exploit and enrich knowledge of threats. The manufacturers play a fundamental role in this point, since they usually introduce and move partners and customers between regions where they expose their best experts and third parties. Cutting-edge manufacturers, who want their products to continue to be leaders, move a whole host of experts and challenges to hackers, to continually test their hardware and software and thus differentiate themselves from the rest of the market.

Point 6 is very healthy because it forces evolution, not just a mere people-outsourcing to be up-to-date and certified. Beyond the classic certifications like ISO 27001, which applies in the Cybersecurity providers 1.0, since the 27001 is too dependent on the scope and how is implemented, but you can almost define a very narrow scope, so ISO 27001 could be completely useless. This explains why new players are emerging in the world of certification of security and managed security providers, as, the LEET company, in the Spanish market, that offers a specific certification on the quality of the managed security provider, and its solvency and guarantee in the services offered. These new players appear on the market to help renew the certifications and excellence of security providers, and separate 1.0 providers from 2.0. Being informed and analyzed is highly recommended, fresh air that the English say, a little fresh air regarding the certifications of MSSP and SOCs that require much more than a certificate to be a CERT or ISO 27.001.

Finally, point 7 is a mandatory requirement to be a service provider 2.0 as the waves of threats, attacks and vulnerabilities are generated as we say in IT, following the sun, around the world and the competitive and operating advantage offered by the Having teams and service centers on 2 or 3 continents is the difference of having

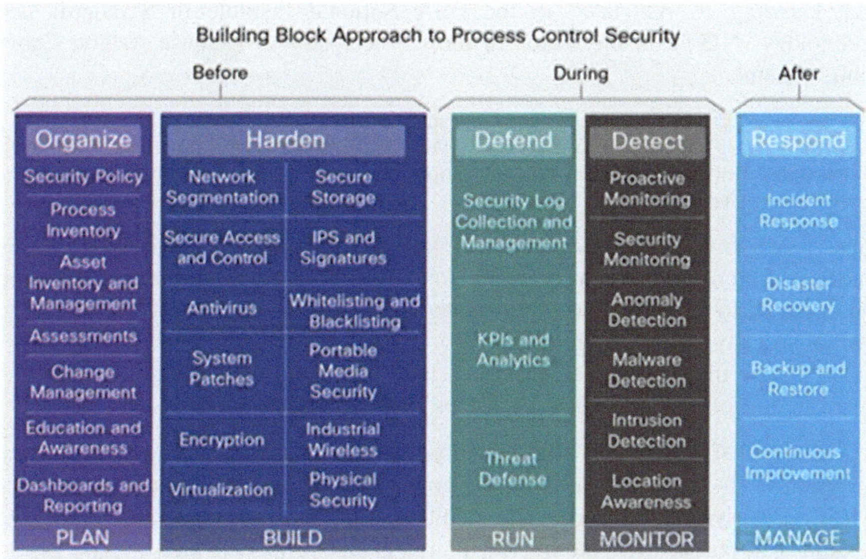

Fig. 4 Advanced Security Services Portfolio for 3.0 Providers

100% of services down or having only 50% because one region suffers and warns the other, so multi-territoriality is a characteristic that Must have obligatorily the providers Cybersecurity 2.0, we must put like a must-have to be supplier services 2.0 (Fig. 4).

Cybersecurity 2.0 Intelligence quality of services 2.0 is based on anticipation, knowledge, experience and infiltration. It is impossible to provide advanced services without a certain level of communication with the dark side, either because you have a first level Tiger-Team, or because you have the networks in the Deep-Internet for that transfer of knowledge.

Intelligence in services is defined by several elements: the first to attack and test your own services continuously, the first attacker and analyzer of your services must be yourself. A good 2.0 vendor submits its vendors and architectures to analysis, criticism and continuous attack, providing intelligence to prevent and anticipate attack vectors before they come to the "bad guys."

Additionally, the intelligence network is necessary to create it from outside anonymously so the so-called Tiger-Teams of teams endowed with advanced security consultants, interact with hackers, crackers and curious several external restless people who enriches with their curiosity and dedication to Highlight as an invasion the knowledge of the service centers that they want to advance this knowledge to their customers to improve their security.

Diversity of threats, diversity of origins, targeted attacks and APTs are elements in which the intelligence or cyber intelligence features of the vendor 2.0. They will mark the difference between proactive defense or being objects of attacks without even knowing it. As stated in the USA National Institute of Standards and Technology (NIST), in the words of the US secretary of Defense Asthon Carter public statement:

> It was our worst fear: a rogue program operating silently on our system, poised to deliver operation plans into the hands of an enemy. Our worst fear: a malicious program (malware) operating silently in our system, remote control to carry out a damage plan in the hands of the enemy (Carter 2011).

These words from the US Secretary of Defense perfectly reflect the reality of the transformation of Cybersecurity in our days and the future: silent threats designed and executed to measure against our systems, designed to dodge our perimeter and internal security barriers, slowly but firm.

These new threats are used for fraud, theft of assets, theft of money, sabotage, terrorism and privacy intrusion, examples such as:

1. Stuxnet (2010): It is the first known worm that spies and reprogram industrial systems,[1] in particular the Supervisory Control and data Acquisitions (SCADA) systems of control and monitoring of processes, being able to affect to critical infrastructures like nuclear power plants, refineries, etc.
2. Flame (2012): Is a malware engine derived from Stuxnet. Flame can be spread to other systems through the local area network (LAN) and through USB memories. It can record audio, screen shots, keystrokes and network traffic. The program also records Skype conversations and can control Bluetooth to try to

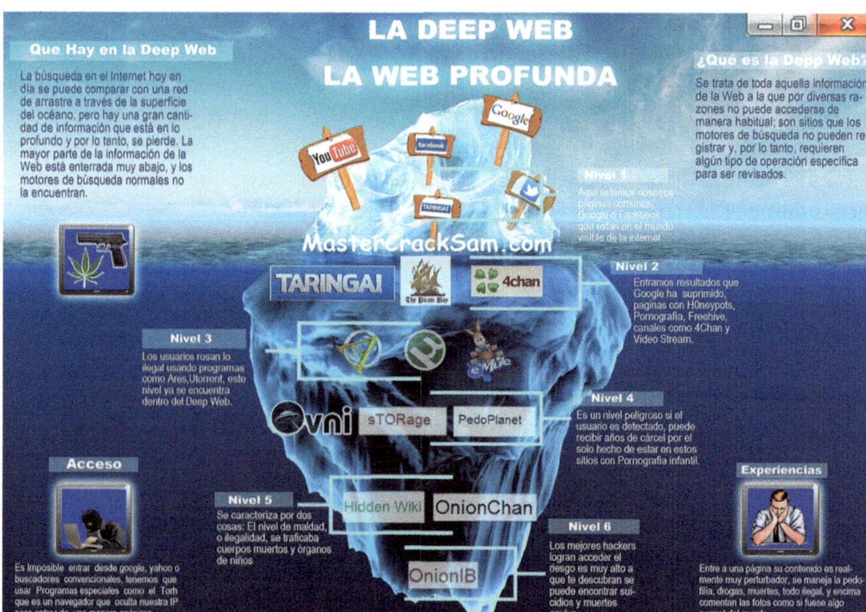

Fig. 5 Public Internet, and Deep Web or Dark Internet Levels and dangers

get information from nearby Bluetooth devices. This data, along with documents stored in the computer, are sent to one or more servers scattered around the world. When it finishes, the program is kept waiting until it receives new instructions from those servers.
3. Operation Aurora (2009–2010): Set of Cyberattacks based on custom Advanced Persistent Threats (APTs), supposedly issued from China against large American companies, as main target Google.
4. RED OCTOBER (Oct 2012): Trojan discovered by Kaspersky in Oct 2012. spied the motives of many European diplomats between 2008 and 2012.

This type of new malware (malicious programs that combine different techniques of infection, attack and distribution), some defined as APT, require a new type of protection services. In many cases they are just the tip of the iceberg of the thousands of APTs that are in production and have not yet been discovered (Fig. 5).

3 Security Services Generation 3.0: Cybersecurity and Intelligence, Strategy, Virtualization and Cloud

We have reached our current environment, the world of advanced services Cloud, mobile, IoT, Smart Cities, SCADA industrial control systems, in short the "internalized world" and digital that brings so many advantages to the end user. But this new environment requires a new type of security provider and new services, in addition to a fundamental change in the mentality of corporate security managers (CSO or CISOS), security cannot be given individually or in isolation, External resources must be used, either from vendors, manufacturers or from third-party sources, but it is IMPOSSIBLE to offer security in a large company in our own days only. And why? Let's see.

The world of Cybersecurity 3.0 is a new scenario that every company should sit down to value in the face of its strategy and rewrite its Master Plan for the next 2 or 3 years, in the current times I would not think of doing a Master Plan more than 3 years with the variability and evolution of technologies and threats is crazy. In fact, many companies with a good sense of strategy are developing a White Paper on Security 3.0 as they have understood that many areas of companies will need a safety guide that helps them to have criteria, and why? Or what can the Marketing department of a large insurance company, or hydrocarbons, or sale of food or textiles need a white guide safety guide? Simply because there is not a single company of any sector that are not doing business over the Internet, or developing your payment and loyalty Apps or forcing your suppliers to interact for invoices and orders via the Internet. None of them has considered Security as the priority for that projects, or a priority in their market strategy, against other more standard objectives for the business as, getting more Customers, loyalty, or increase markets or growth of business in general.

The Cybersecurity 3.0 born in this new ultra-digitized world, combines seven determining factors that give us the keys to survive in the new ecosystem:

1. The attacking element is dynamic and has intelligence, adapts to the security measures of the Client and the changes of the environment, to remain hidden and operative in the maximum number of systems, for the maximum possible time.
2. Combining elements of disparate technological nature that interact autonomously and in real time, such as SCADA systems and industrial systems (navigation area, satellite control systems, drones, car driving systems), M2M and IoT, millions of teams communicating and acting with absolutely no human intervention and real-time Smart Cities, cities that interact with citizens' systems independently and automatically.
3. Developing new applications per million daily, the program code element becomes a key element in the new security field, since in both the mobile and fixed worlds, Mobile Apps generate millions of new programs appearing and Combining daily and instantly installing on billions of devices. This phenomenon has triggered the number of programs where safety is not a priority.
4. The Cloud and IoT eliminate the perimeters of security. There are no boundaries between systems, companies and users, and the field of action, impact and attack of new threats is virtually infinite, a whole "greenfield" for attacks. The adoption of the Cloud by non-technological departments such as marketing, HR, financial, shopping, etc. generates exposure of areas that security departments are not accustomed to face, nor are there physical technological measures (on premise) to deploy. You can protect and monitor Cloud services only with Cloud security services.
5. The volume of information and data to be controlled grows exponentially. It multiplies the volume data generated by the attacks and extends over time, days, or months. The huge number of new applications and systems cause an avalanche of generated data and companies are absolutely inefficient in collecting all necessary security information. In addition, many of the apps do not generate logs, at all, not to mention that many Cloud services do not have logs or do not allow access to them. Collecting and appreciating all this information is key in the services of Cybersecurity 3.0.
6. The value of information and virtualization of currency and economic transactions, combined with virtual digital currencies, promotes and reinforces cybercrime and cyber fraud, since they can capitalize on successes immediately, anonymously and unpunished. Also information is money; so you do not need to steal money so that the fraud or attack is successful. Habits, personal information, location, tastes, relationships and in general any information of users and companies mean money.
7. Intelligence, collaboration, anticipation and real-time action are the keys to effective protection. These are the characteristics that the systems offered by the Managed Security Providers (MSSP), by manufacturers of software and

hardware of security 3.0, or by the combination of all or some of them, depend the effectiveness of our protection and of the security of our assets and people and their information. It is fundamental to belong to common intelligence networks, to share knowledge and threats and to deploy.

These factors must be taken into account and combined in the 3.0 services and products, and they affect separately but in combination the security deployment. It also gives us a simple guide to identify the structure of the portfolio of Cybersecurity 3.0, which should include at least:

Group 1—Cloud security services (EYE! They are cloud security services not cloud services with security, not to confuse)
Group 2—Application protection and analysis services and code
Group 3—SIEM and Big Data Services
Group 4—Cyber Intelligence Services.

Of course, the traditional services of integration and deployment of technologies can be included in the portfolio, and managed services of technology on premise, but being well known and included in the generations 1.0 and 2.0, we do not consider them. While the complexity and depth of the new 3.0 service families is generating specific specialist providers and it is becoming increasingly difficult to be a general provider that can cover the four groups of new services. The new cybersecurity providers (MSSPs combine their own and third-party solutions to cover the four fields in a comprehensive and coherent way, and also manage to integrate different services and technologies in a simple way. This is the future that MSSP 3.0 will play.

4 Conclusion

The seven factors described in the previous section configure a present that leads us to a future of changing the cybersecurity environments without turning back. Unfortunately, most companies (large and SME) ignore or belittle this evolution, which is already costing billions in image damage, theft, fraud and loss of information: just remember the cases of Sony or the most recent of Ashley Madison, or the latest XSS attack on Salesforce (Siluk 2015), although it also happens often in Spain and Latin America.

The adoption of the Cloud in business processes is accelerating the disappearance of perimeters and companies require new services and products capable of adapting to this new dynamic and virtual environment. There are cases in which large corporations are developing advanced capabilities to protect themselves and it could be the case that even in the near future we could see how these corporations specialized in SCADA or Financial services also become, why not, in Cybersecurity 3.0 vendors for third parties leveraging their internal capabilities and technologies.

The future is open, and the only thing certain is that with the exponential multiplication of the Cloud, the IoT and mobile applications, we force security vendors and manufacturers to move forward even faster. We are not going to save the security by design that many developers do not apply because the growth of the new environments is exponential. Consequently, we need the services, products and suppliers of cybersecurity to evolve at the same speed, because the world of the perimeter controlled and located, will not return.

References

Carter A (2011) Preliminary Cybersecurity Framework. NIST (Feb 2011)

Siluk S (2015) SalesForce closes door to hack attacks. CIO Today, 12 Aug 2015. Retrieved from: http://www.cio-today.com/article/index.php?story_id=101003937V84

Author Biography

Juan Miguel Velasco is the founder and current CEO of AIUKEN SOLUTIONS, a Security and Cloud Services Company that operates in Spain, Andorra, Puerto Rico, USA, Morocco and Chile. He is also the VP of EuroCloud Spain, President of the Eurocloud Cybersecurity Commission and Secretary General of the Advisory Board Committee at iSMS Forum Spain. Previously, Mr. Velasco held diverse executive roles at various Technology and Telecoms Companies, such as Senior VP of Business Development at Buguroo Security (2012), and was Executive at Telefonica Group since 1999, acting as Director for the Security Services Division at Telefonica Enterprises Spain (2006–2012); Deputy Director for Security Services and Shared Platforms at Telefonica Empresas (2004–2006); Director of Infrastructures Planning and Architectures Strategy (2003–2004) and Deputy Director of Project Engineering and Security Services at Telefonica Data Spain (2002–2003). Also Chief Operations Officer and Chief Technology Officer at ACE (Electronic Certification Agency) company created by Telefonica DataCorp and the Spanish Association of Banks 4B, CECA and SERMEPA-Visa Spain (1999–2002).

Collaboration of Private Investigation with Public Institutions Within the Spanish Cybersecurity Strategy. How Private Investigation Gathers Proof on Cyber Delinquency

Francisco José Cesteros

Abstract There are several reasons to consider cybersecurity risks for companies and personal information. Companies can be attacked and are exposed to hacking and comments that affect their reputation, their brand and the security of their employees. The assets of the organization are the employees, connected computing devices, external users, services and applications, social networks, communications systems and all of the information transmitted and/or stored in the cyber environment. Personal and familiar information is also exposed as children use social networks, we share the computer and cloud at home and the profiles we use are not always secured, showing a lot of information that reveals aspects of our private life, without talking about the Wi-Fi penetration and information theft (pictures, bank accounts, documents, etc.). The Spanish National Cybersecurity Strategy is the framework of an integrated model based on involvement, coordination and harmonization of all stakeholders and state resources in public-private partnerships, along with the participation of citizens. In order to do this, a strong coordination of the various government agencies, as well as adequate public-private cooperation initiatives to be able to reconcile and promote the exchange of information are required. Private investigation professionals, working within the law, and using technology, forensic methodologies, and cyber intelligence procedures help in finding the offenders and getting the proofs for trials and lawsuits.

Keywords Private investigation · Cybersecurity · Cyberbullying · Grooming · Social network · Homeland security

Submitted: 30.07.2016; Accepted: 6.9.16.

F.J. Cesteros (✉)
Cuzco Detectives, Madrid, Spain
e-mail: fjcesteros@cuzcodetectives.com

J.M. Ramírez and L.A. García-Segura (eds.), *Cyberspace*, Advanced Sciences and Technologies for Security Applications, DOI 10.1007/978-3-319-54975-0_10

1 Introduction

There have always been attackers and defenders in the disputes between humans, linked to survival, to conquer areas because of their natural resources, minerals, energy, water, agriculture, industry expansion and of course, political and ideological messages used as propaganda.

Traditionally, security and defense have been based on physicality: materials, weapons, and landscape to fight; than on a virtual field of operations. But the new times the field of operations has moved from the ground to the virtual ground, from the gunpowder to the bit and byte, and from face to face fighting to using anonymity as a weapon. Thus we require new ways of thinking, understanding technologies, and vulnerabilities, and creating a security framework where public and private institutions collaborate in a general interest.

On December 5th 2013, the Spanish Government launched its National Cybersecurity Strategy, being the first time it made public awareness about the threats and risks of the cyber environment. This clearly states that there must be a shared collaboration and coordination between public and private security, as it is also mentioned in the National Security Law, which regulates and defines that

> National Security should be considered a state, regional and local shared by the various Administrations objective, the constitutional bodies, especially the Parliament, the private sector and civil society, within projects of international organizations of which we are part. The reality shows that the challenges for National Security that affect society are sometimes highly complex, which goes beyond the boundaries of traditional categories such as defense, public security, foreign affairs and intelligence, and more recently incorporated into security concerns, such as environment, energy, transport, cyberspace and economic stability (National Security Law 2015, page 2, on my own translation).

Spain thus is concerned about security and cybersecurity as a whole, moving into actions such as: laws, cooperation, collaboration, awareness, a National Security Department .... This also requires the effort of different security institutions, private companies, intelligence services, changes in regulations, acquisition of technical knowledge, etc. Especially pressing is the need to response to the advance use of digital networks by non-state actors for criminal purposes or associated with terrorism and organized crime.

2 A Review of the National Security Law and National Cybersecurity Strategy

Besides the previously introduced National Security Law (2015) and the National Cybersecurity Strategy (2013), there is also a Private Security Law (2014). Let us review some important messages and points of these documents.

2.1 *National Security Law (2015)*

The objective of this law was to establish the prevention and response framework to solve problems regarding the national security and coordination between the different levels and institutions of the public administration, and to facilitate the participation in national security matters of private companies and the civil society.

The Law is divided into five titles. The preliminary one, in addition to the provisions relating to its object and scope, provides definitions and general principles underlying the concept of Homeland Security, such as state policy, culture of national security, cooperation with the autonomous communities, private partnership, key components, as well as areas of special interest and obligations.

Title I details what the competent bodies of the National Security and what powers are assigned in this matter.

Title II is devoted to the creation and definition of a National Security System and a National Security Council, with its functions, and organization.

Title III regulates the management of crisis, the general operating framework of the National Security System and establishes definitions and skills in this area.

Finally, Title IV regulates the contribution of resources to the National Security:

> The National Security Strategy is the strategic policy framework of National Security Policy. It contains the analysis of the strategic environment, the specific risks and threats to the security of Spain and defines the lines of strategic action in each policy area and promotes resource optimization. (National Security Law 2015; own translation)

Especially interesting is article 7:
Article 7. Private entities.

> 1. Private entities, because due to certain circumstances, and in all cases, when they are operators of essential services and critical infrastructure that may affect national security, should collaborate with the government. The government shall establish the mechanisms and forms of this collaboration.
> 2. The Government, in coordination with the Autonomous Communities, will establish channels that encourage private sector participation in the formulation and implementation of National Security Policy. (National Security Law 2015; own translation)

Article 8 involves the participation of the civil society:

> Article 8. Citizen participation in National Security. The Government, in coordination with the Autonomous Communities, will establish mechanisms to facilitate the participation of civil society and its organizations in the formulation and implementation of national security policy. (National Security Law 2015; own translation)

Article 9.2 goes directly to the core:

> Article 9. Key components of National Security. (…) 2. Intelligence and Information Services of the State, according to the range of its powers, will permanently support the National Security System, providing evidence, information, analysis, studies and proposals necessary to prevent and detect risks and threats and contribute to their neutralization. (National Security Law 2015; own translation)

And article 10 defines the areas of special interest:

> Article 10. Areas of special interest of National Security. Areas of special interest of National Security shall be deemed those who require specific attention and are basic in preserving the rights and freedoms and well-being of citizens, and to ensure the supply of essential services and resources. For the purposes of this law, they will be, among others, cybersecurity, economic and financial security, maritime safety, airspace security as well as outer space, energy security, health security and preservation of the environment. (National Security Law 2015; own translation)

Again, the private sector is mentioned in article 27.5:

> Article 27. The contribution of resources to the National Security System. (…) 5. The private sector will participate in the contribution of resources to National Safety. (National Security Law 2015; own translation)

So, it is clear that the Spanish public administration has developed this law thinking of society and private companies as a part of the National Security Strategy, transforming it into a space of collaboration and mutual commitment.

In 2012, three years before the aforementioned law, the Department of Homeland Security was created. It started to work on the National Security Strategy, including cybersecurity and regulations. A part of the Department, the Office of Security and Information Technologies, serves as the advisory body to the Prime Minister on National Security and assumes the functions of Technical Secretariat and permanent working body of the National Security Council.

We can note that Spain has been really focused on organizational structures, regulations, strategies and actions to guarantee a safe place for both citizens and companies.

It must also be underlined the value of Spanish legislation in protection against the security of digital information networks, as well as the legal reforms undertaken to classify new crimes related to cybercrime, cyberterrorism and aligned with the European Union Guidelines (ENISA 2012).

2.2 *National Cybersecurity Strategy (2013)*

Since the use of Information and Communications Technologies is integrated into our daily life, there are serious risks and threats that may affect national security and critical infrastructures, private companies, families and social coexistence alike. Accordingly, there are several factors that allow the proliferation of crimes:

1. The profitability in economic terms.
2. The profitability in political or ideological terms.
3. The low cost of the tools used for attacks.
4. Anonymity.
5. The profiles of attackers who exploit technological vulnerabilities in order to gather information, remove high-value assets and threaten basic services.

What the Spanish National Cybersecurity Strategy stands for?

> National Cybersecurity Strategy is the framework of an integrated model based on the involvement, coordination and harmonization of all actors and state resources in public-private partnership, and citizen participation. Also, an essential part of this model is given to the transnational nature of cybersecurity cooperation with the European Union and other international agencies or regional level with jurisdiction in the matter. (National Cybersecurity Strategy 2013; own translation)

The strategy consists of five chapters.

Chapter I, entitled Cyberspace and Safety, lays the characteristics that define cyberspace, the opportunities and implications of it, all from the viewpoint of safety.

Chapter II establishes the purpose and the guiding principles of cybersecurity in Spain, through an integrative vision involving the coordination of government, the private sector, and citizens. It channels international initiatives on the matter, regarding domestic and international law, and in line with other national and international strategic documents.

In Chapter III, the strategy addresses, with an increasing level of detail, the objectives of cybersecurity. The global objective is to make Spain safer in the use of Information Systems and Telecommunications, strengthening prevention capabilities, defense, detection, analysis, research, recovery, and response to cyberattacks. The Cybersecurity National Policy should serve for this purpose.

Chapter IV describes the action lines of National Cybersecurity, linked to the established objectives in Chapter III.

Chapter V is dedicated to cybersecurity in the National Security System and establishes the structure serving this purpose.

As a result, we can summarize it by introducing what the Chapter I says: (Presidencia de Gobierno, Gobierno de España 2013), (own translation):

> The development of Information and Communications Technologies has generated a new space where the speed and ease use of information exchange and Communications have eliminated barriers of distance and time. (National Cybersecurity Strategy 2013; own translation)

Cybersecurity thus has as a global goal to make Spain safer in the use of Information Systems and Telecommunications, strengthening the capacities of prevention, defense, detection and response to cyberattacks. The growth of the country depends on the safety and reliability of our cybersecurity.

The following points are addressed in the cybersecurity strategy:

1. Foreign states
2. Technical issues or attacks
3. Hacking
4. Organized crime
5. Terrorism
6. Crimes

7. Natural catastrophes
8. Conflicts
9. Sabotage
10. Espionage.

2.3 *Private Security Law (2014)*

What does the collaboration of public and private institutions mean regarding security and cybersecurity?

In our European environment, private security has become a real actor of global and national security policies, being increasingly considered an indispensable part of the set of measures for the protection of society, along with defending the rights and legitimate interests of citizens.

The Spanish Private Security Law (2014) is structured in six titles:

Title I reflects one of the key ideas that inspired the drafting of the law: the coordination and collaboration of private security services and the security forces, with the sole aim of improving public security, through the exchange of information always within legal guarantees.

Title II gives legal status to some provisions dedicated to the regulation of security companies and detective offices, or both records, which stay unified in the new National Register of Private Security.

Title III regulates issues previously dispersed in different norms, such as those relating to the functions of most of the security personnel.

Title IV regulates for the first time a series of security measures and specifies the provision of the main security services (monitoring and protection, protection staff, storage and transportation security, and private investigation).

Title V regulates, also for the first time, the control and inspection of institutions, staff and security measures and as well as their obligation of cooperation.

Title VI gives solution to some of the major shortcomings of the previous legislation relating to the sanctions regime.

Furthermore, we find relevant to mention the following articles of the law:

> Article 16. Coordination and participation. 1. The Ministry of the Interior or, where appropriate, the competent regional authority shall take organizational measures that are appropriate to ensure coordination of private security services with those of the security forces. (Private Security Law 2014; own translation)

The activity of private investigation is allowed to private detectives and private investigation agencies as referred in the law:

> "Article 48. Private investigation services. 1. The services of private investigation by private detectives consist of conducting inquiries that are necessary for the collection and input on behalf of legitimate third parties for information and evidence of conduct or private events related to the following aspects:

(a) Relating to economic, labor, commercial, financial sector and in general, personal, family or social life, excluding what takes place in homes or reserved places.
(b) Obtaining information tending to ensure the normal development of activities taking place at fairs, hotels, exhibitions, shows, events, conventions, department stores, public places of grand assembly or similar areas.
(c) Conducting investigations and obtaining information and evidence concerning only prosecutable crimes on behalf of persons authorized in criminal proceedings" (Private Security Law 2014; own translation).

It can be concluded therefore that Spain is moving forward into the warfare phase of cybersecurity because the minds and behind the regulatory changes both public and private, are alerted and working on it.

2.4 *Cyberdefense Joint Command (Ministry of Defense)*

Another point that demonstrates the real focus of Spanish Government, aligned with the European Strategy is the creation in 2013 of the Cyberdefense Joint Command (MCCD, following the Spanish acronyms).

> The Cyberdefense Joint Command is part of the operating structure, subordinate to the Chief of Staff of Defense (Jefe Estado Mayor de la Defensa—JEMAD), responsible for planning and implementing actions relating to cyber defense in networks and information systems and telecommunications (Ministry of Defense) or other institutions that could be entrusted, as well as to contribute to the proper response in cyberspace to threats or attacks that may affect national defense.

The previous relevant dates related to this commitment were:

1. 2001, January—approval by the Chief of Staff of Defense of the "Vision of the Military Cyberdefense".
2. 2011, July—approval by the Chief of Staff of Defense of the "Military Cyberdefense Concept".
3. 2012, July—approval by the Chief of Staff of Defense of the "Action Plan for Obtaining Military Cyberdefense Capability".
4. 2013, February—Defense Minister promulgates the "Ministerial Order 10/2013, establishing the Cyberdefense Joint Command creation".

2.5 *Summary*

The Spanish Government has showed a clear interest in commitment and innovation regarding one of the biggest issues that we actually face, cybercrimes, creating institutions and launching changes in the regulations. These modifications include the sophistication of technologies, the collaboration between public and private institutions, along with the economic resources required to be aligned with European Laws and common defense and strategy, as well as an increase in the budget for the local institutions with responsibility in the fight against crimes and cybercrimes (Ministry of Defense, Homeland Security Ministry, Intelligence Service, Ministry of Justice and Ministry of Public Administration).

Consequently, we are now better prepared to understand the situations and the weapons, facing them day by day. The following section is dedicate to analyze how the private investigation is doing.

3 Private Investigation

In most of the cases, what customers expect from the private investigation agencies is to find out who is behind a fraud or bullying, or who is penetrating their information systems. Companies usually contact us because their systems have been attacked, or they understand that their personnel are under some kind of pressure from external people.

The methodologies to investigate what is going on are different, depending on the circumstances. Which are the main situations we are consulted on?

1. Company's employee and social networks
2. System information attack
3. Bullying on employees
4. WiFi penetration
5. Information theft
6. Online fraud
7. Identity theft
8. Bullying on children.

3.1 Company's Employee and Social Networks

Due to the recent economic crisis in Spain and the explosion of social networks and devices, anyone can generate a profile in a social network and use it to talk about anything, be it real or false statements.

Companies are investing more and more money on social networks and social responsibility image in order to grow, but all the money they invest could be wasted if they receive negative messages on these social networks. The consequences may be economic damage, loss of customers and a false image sent to the market. Because of these tremendous losses, companies require a full investigation regarding what messages are damaging their image, where are they coming from and acquiring real evidence to start a trial.

At this point is where private investigation starts looking into the social networks and matching the messages of each profile with possible employees. Experience, methodology and tools for the analysis are very important. IP addresses, profiles, contacts, friends, followers, pictures, etc. and the use of graphical tools let us know and understand the relationship between messages and people.

Although it is quite complex and requires time, sooner or later the results arrive and once the matches appear, the next step is talking with the employee and solving the situation.

Nonetheless, cyber comments or the use of social networks are more and more extended because people can hide behind a false identity, and employees can spread comments against a company. This is one of the most common attacks coming from inside the company. It requires time, methodology, tools and knowledge about technical and social networks.

The OSINT (Open Source Intelligence) methodology is critical for getting the evidence required for sorting results and especially relevant for linking contacts, messages, and registers in different social platforms.

With the proof of harm at hand, the forensic track of the investigation is evidence as well and, if it goes to trial, the methodology will be another asset for the investigation.

In summary, when investigating cases of social networks, methodology, technical knowledge, tracking and forensically probed steps are required to catch the offender and then clean the image of the companies and improve their reputation.

3.2 System Information Attack

Another kind of cybercrime is the one related with system information, including networks and storage. In Spain, it is identified as a crime, even in the cases involving probing a system for their own knowledge and capabilities. The other typified reasons are for damaging the availability of the web, for information robbery and for acquiring information and selling it the black market.

What is clear is that no one attacks an information system for nothing. There are different interests for accessing information systems, networks and information stored on systems. Not all of the attacks start with a police investigation. Companies often want to contract with a specific private investigation agency, track the attack, and get a forensic analysis and evidence of the crime. Then, once the attack has been valued, security and justice officials are contacted, or simply, the attack is hidden. In fact, not all the attacks need to be shown to the public because talking about them can create alarms and harm the image of the company, but that doesn't mean that they didn't exist.

Zero-day vulnerabilities are not often used, but when they do appear they need to be known in order to solve them as soon as possible. Attacks show vulnerabilities, flaws on the architecture design, lack of updates and more often than expected, flaws in the management process without a proper backup, without a properly probed procedure and lack of knowledge of the technicians. And of course, relating directly to the company, a lack of investment in securing systems, networks, procedures and training of their technicians. In some cases, the investigation shows just an intrusion and beefing up, updating systems and eliminating vulnerabilities can stop more attacks.

In summary, all the different motivations for cyber offenders are related to information system vulnerabilities, architecture design and investment in security. Companies need to understand that investment is security is part of their corporate image, credibility, and profit.

It is not possible to grow without paying attention to the security design and technical issues that appear while the company's web and information systems are 24×7 online; meanwhile the sales, information, products, services, documentation, provider's access, and so on depend on the availability of the systems and networks, the secure infrastructure and the investment on updating knowledge and best practices.

Given that private investigations start once the intrusion has been committed, we recommend using a preventive approach to the cybersecurity risks and investing in security.

3.3 Bullying on Employees

Another situation we fight and investigate is the bullying that some employees suffer from Internet users and from other companies using their network facilities and social links.

We are usually contacted by the company or by the besieged people in order to find the stalker, and the reasons and goals behind him. From our experience, stalkers investigate the company and look for the most interesting people to get access to information. Sometimes they focus on executives as they manage information and money; also the financial director or the sales director are usually interesting people. Stalkers get in contact with them by using social engineering to access their e-mails or phones. The secretary's phone is not always a barrier, and they can get the contact. Once they have contacted the executives, they start off with photos of them (their families or with ones gathered from private parties). Information is available for the stalkers. They manage the data and the time.

Surveillance, and counter surveillance, of the executive and his family, college and life style is part of the process. If the executive is single, party photos are relevant as they are used to have some interaction with the other sex and some pictures are taken, they are then exposed out of context, including alcohol consumption and Photoshop manipulation.

Social engineering is also important when a stalker wants to approach the executive. Money is on the table and stolen pictures inside the office or private parties are part of the game. So, there are many situations where stalkers use social engineering and technology to approach the victim and get access to information.

What stalkers are looking for? Their goal is getting business information on sales margins, providers, etc. from the corporate database, in order to have access to customer's profiles or paying money for specific information, mobile phones, addresses, pictures, etc. In other situations they want to destroy the corporate image as well as to destroy the reputation of the executive because of some internal affairs or external situations involving certain habits and lifestyles.

This is not the only way of acting. Another one is based on a technical people approach or administrative people approach. Stalkers pay people or extort them in order to access information. Our goal is to discover the stalker, get the evidence and stop any bullying by being part of the legal process.

Behind the bullying, there is money, industrial espionage, channel information, sales margins, access to providers and reputation destruction. Sometimes the blackmail is due to personal affairs, infidelities or gamblers that discover the dark side of people and expose it to society or their colleagues.

3.4 Wi-Fi Penetration

Given the technology evolution, we all are dependent on WiFi. Companies' networks usually use cable connection, but in some specific areas, as an assett for visitors, Wi-Fi connection and accessibility to the internet are permitted.

Normally the network administration understands that Wi-Fi is a hole where information tracking can occur. Therefore they try to protect it. But sometimes, due to the requirements of a meeting, a conference or a presentation, the administrator can install Wi-Fi access without all the care it requires, and here is where problems come. The coverage of the Wi-Fi, the repeaters, the password and so on are part of the network's administrator's work.

But there is another situation where risks occur, and that is personal Wi-Fi at home where you are very exposed to Wi-Fi penetration and information theft. Our investigations start once the information has been taken and alerts are high. The Wi-Fi penetration at home and in companies is investigated and a forensic analysis is made in order to determine the stolen asset and the impact of the penetration. This forensic analysis determines the impact of the vulnerability, establishing the best way of working and getting the proof of the possible illegal information theft.

Families usually share tablets and laptops at home, and they do not secure each profile, using the same one to get access the internet and hard disk. The consequences are that children can access to specific and confidential information in the files of the whole members of the family and, if the network security is weak, it can also be accessed by the intruder. This lack of security is a risk as parents get their work files at home.

The entire family needs to know that accessing the internet from public sites has not only advantages but also risks. The best practice you can start at home Children, along with their ambition and expectation for "always being connected" are the first step we need to be worried. They have to be educated on security awareness, on being cautious about their own visibility, images and personal information as same as their family information.

Wi-Fi penetration is part of the forensic analysis and investigation to discover theft of information and not just a penetration in order for access the internet. We have to worry about the security of our wireless connection in public spaces and the information we store in our devices. Private investigation helps finding the intruder and submitting them in a legal process, but remember this is done once the illegal penetration has been made and perhaps the theft of private information and data.

3.5 On-line Fraud

Interesting risk when the technology evolves. The use of devices like the "Pineapple" which captures the information of mobile devices that are in its scope and have their Wi-Fi or Bluetooth activated.

But this is not the only way for gathering information. As we have seen before, Wi-Fi penetration and bullying are other ways of capturing or getting specific data. The use of their laptop, tablet or mobile phone for online shopping, is part of a new trend of acquiring clothes, devices, etc.

Since not all the webpages are using secure technology for their online payments through credit cards, buyers need to be sure about the different methods of payment, and that webpages do not always offer the buyer a secure connection, and this is where intruders and hackers go to acquire the data. There are some specific procedures using Western Union, PayPal, specific Credit Cards, bitcoins, or another ones that allow the payment and of course the fraud.

Private investigators' experience is based on the knowledge of the technologies and the procedures that organized crime use for acquiring stolen credit card numbers. For instance, the forensic analysis of the system along with the web profile is where the private investigation goes for getting the evidence. This starts once the crime has been committed, and gathering proof is the first step for the complaint.

Online fraud involves theft of data, and the subsequent use of it along with payment through the stolen credit card information of another person. Criminals and organized crime also know what Spanish law says about these crimes, specially the 400 euros barrier: if the purchase is below this amount, then the crime carries less punishment. We recommend the use of a specific credit card for online shopping, issued to a specific bank account with limited funds and once purchases are made, to verify that the payment method is a secure connection, and there is a secure method of payment.

Private investigations usually track the transaction and then prepare a report, collaborating with him the customer on getting the evidence and moving the legal procedure. But unfortunately we do not have the capability of getting the money back. Our experience is based on the knowledge of the technologies and the procedures that organized crime use for acquiring stolen credit card numbers.

3.6 Identity Theft

Due to the expansion of social networks and the current need for communication that people have, the profiles on the internet are actually one of the most interesting data that criminals want to get and want to use. Fraud, social conversations, and blackmail, are used as part of identity theft, allowing criminals to use another person's identity for their fraud, blackmails, etc.

A person's identity is protected by law, including the legal use of a pseudonym or an alias or coverage for the investigation. It is also protected by law the identity of the undercover agent, authorized by a judge and of course under monitoring and continuous report, but this legal creation, part of the Police's investigation procedure, is not what we are talking about. Apart from the above cases, identity theft (a real identity) is a crime, and consequently it is punishable by the law.

Identity theft or usurpation of identity is pursued by criminals in order to earn money and/or to chat in blogs or social networks. They use it as coverage for their political claims, hiding and writing what they want and, of course, they use it also for blackmails.

Recently, an article about blackmail was published in a digital newspaper (El Mundo 2016), showing how clear impact it has on the reputation of the victim. As in other other crimes, the main objective of the criminal is payment.

Identity theft is also used for other crimes, such as getting e-mails, credit card numbers, and purchasing or grooming (behaviors and actions deliberately undertaken by an adult in order to gain the friendship of children, creating an emotional connection with them) for example. Another risk to be aware is how do they get it? Sometimes they send a spam e-mail to unwary people that still believe in prizes, bill imitation, PDFs files, etc.; other times they install key loggers (hardware or software) on public computers.

Again and again, the best way for protecting yourself from these crimes is to be secure, to be aware and to be alerted in order to distrust about any easy methods of earning money.

The private investigation carries out the forensic analysis of the systems, tracks the identity, the profiles and prepares the report for Police. The final step is the complaint.

Since the Spanish Police resources are not enough for every single case of identity theft, they require the collaboration of private investigators, based on a demand of the person that suffered the robbery.

Sometimes it is easy to find the criminals because of the tracks they leave behind, but in most of the cases this is an expensive and complex investigative process because criminals hide their tracks and themselves. But remember that no crime is alone in the scope of organized crime. Most of them are part of a more complex strategy, which use different and interconnected ways for committing crimes.

The investigation starts in one way and usually turns to other ways, so the requirements of the investigation have to be focused on a proper methodology. Technical knowledge, methodology, focus and experience are the indicators for choosing a private investigation agency.

3.7 Cyberbullying and Grooming on Children

This crime is not only coming from organized groups but also from children, teenagers, and even adults. When involves only children or teenagers, it is known as cyberbullying; and when it involves adults approaching young people, it is known as grooming, Cyberbullying is more than a children's game. It is a crime, and as such it should be pursued. Children should understand that recording and linking content to social networks has consequences for stalkers and consequences for the victims. And grooming is another serious crime. Children should be alerted of any kind of contacts in order them to be protected and secure.

Cyberbullying is focused specially on schools and parents, due to news reports that are constantly appearing in the press (we can search in Google for bullying news), the level of alert and awareness is high.

The best tool for teenagers and children is their mobile cameras. They use it for everything, including holidays, selfies, parties and, of course, cyberbullying. Teenager's problem is that they ignore what the law says about any civil or criminal offenses. They live thinking that there is no illegal action behind their actions. Therefore, we can get inside their world, characterized by their lack of legal knowledge and the use these technologies for everything.

Nonetheless, since the ignorance of the Law does not exempt one from compliance with it, the advance use of technology and the intensive use of its new characteristics should be accompanied by the education in terms of legal knowledge and the respect the legality of their actions, the proper use of technology, the awareness of cybercrimes, a secure cyber world of everyone and the rational use of the capabilities of the devices.

Grooming is another big challenge. Here, adults are part of the problem, and the legality of the actions and their responsibilities. Grooming is growing, and young people hide it from their parents based on the fear of punishment. Young people have to learn to fight against them and acquire evidence of this crime, understanding that behind an unknown person could be a stalker.

From the private investigation perspective, parents come to us because they have detected alerts in the behavior of their children, and want us to investigate who is bullying their children, to do surveillance, to understand the behavior of the children, and to know with whom they are sharing their time and parties.

One of the best things private investigation can do is educating schools and parents on detection, taking some actions to alert children, arriving before the crime, preventing it.

Yes, prevention is great because young people and teenagers are victims, and the consequences may be very serious and hard to forget. And education is an important part of the prevention system. The private investigation collective uses the education in schools and associations to raise awareness about cyberbullying and grooming; and, with the help and support of the Police, to build a preventive environment and culture of security.

Public and private security must work together to defend children and protect their environment, letting them live securely, being aware of these situations in order to avoid them.

4 Conclusion

In conclusion, we have reviewed what the Spanish government is doing to move up the security and the legal framework as quickly as possible; creating institutions for cybersecurity investigation, analysis, methodology, and procedure; and its understanding, and execution. It has shown a big capability to invest in the cyber phenomenon through people, courses, procedures, frameworks, institutions, and laws that describe how to organize each of the issues that appear in relation to cybercrimes.

Public and private securities are working together under these guidelines, designing proper legal and functional frameworks. But the resources of public security are not enough to properly fight cybercrimes, because most of the investigations fall under the private investigation sector, affecting a private person or a private company.

This chapter has offered a general approach to some of the crimes we face with the use of technology. Unfortunately, we can investigate some of them only when the crime has been committed and then, in a reactive procedure, instead of from a preventive one: we work on forensic analysis, discipline and methodology to recover the evidence and prepare the report for the subsequent claim and lawsuit.

But it is more important that parents and general managers work on security prevention. And the good news is that working on the prevention is also possible at the schools and parents associations, in order to arrive to prevent crimes before they happen, including cyberbullying and grooming. The legal framework and the prevention model involve policemen and private investigators that speak about the cyber world, the crimes, and how to prevent them.

There are some specific actions that private corporations should be aware of, such as industrial property theft, network security, certain personal habits of directive personnel. Private investigators can help corporations in finding holes in their security, as well as to build a barrier against crime.

Families, on the other hand, also need to understand that the information about their holidays, pictures, travels and habits are part of their private life. Children need to be aware about what they share and what information is private, and about the best use of the technologies and the consequences of sharing things in social networks. None should share profiles and storage devices like pen drives, because the risks are high, and the consequences are unmeasurable.

Private investigators help families and corporations, transform safety in a value and an asset for you.

References

El Mundo (2016) O pagas, o te convierto en un pedófilo en Facebook. El Mundo, 25 Jul 2016. Retrieved from http://www.elmundo.es/cronica/2016/07/25/5794821fe5fdea450b8b45b3.html

European Union Cybersecurity Strategy (2012) ENISA, Gobierno de España. Ministry of Defence (2013). Mando Conjunto de Ciberdefensa (Cyberdefense Joint Command). Ministry of Defence, Ministerial Order, Sept 2013. Retrieved from http://www.emad.mde.es/CIBERDEFENSA/

Gobierno de España. Presidencia de Gobierno (2013) Estrategia de Ciberseguridad Nacional (National Cybersecurity Strategy). Retrieved from http://www.dsn.gob.es/es/estrategias-publicaciones/estrategias/estrategia-ciberseguridad-nacional

Jefatura del Estado español (2014) Ley 5/2014, de 4 de abril, de Seguridad Privada (Private Security Law) Boletín Oficial del Estado (83). Retrieved from https://www.boe.es/diario_boe/txt.php?id=BOE-A-2014-3649

Jefatura del Estado español (2015) Ley de Seguridad Nacional (National Security Law). Boletín Oficial del Estado (233). Retrieved from http://www.dsn.gob.es/es/sistema-seguridad-nacional/ley-seguridad-nacional

Author Biography

Francisco José Cesteros is an industrial engineer from the ICAI, and has several diplomas on National Defence and in Cybersecurity by the Spanish Defence Ministery. Professional Expert in Information Systems, Private Investigations, Security and Defence. Private detective; Master in Business Administration; Master in Peace, Security and Defence. Diploma in National Defence and in Cybersecurity by CESEDEN, Judicial expert on engineering, security, graphology, documentoscopy and scientific police. Certified instructor in Lie Detection by Dr. Paul Ekman. Currently working in the field of IT, Research on Security/Defense, with over 25 years of experience and providing capacity, knowledge, methodology, team management and focus on results. At the present, he works as Intelligence Manager of CUZCO DETECTIVES.

Psychosociological Characteristics of Cybercrime

Juan Carlos Fernández-Rodríguez and Fernando Miralles-Muñoz

Abstract This chapter analyses the most important concepts and characteristics about cybercrime, cyberterrorism and cyberwar. For the first two concepts, data are provided on profiles and on the different types of people who engage in this kind of criminal behaviour. In the case of cyberwar, we present the most relevant data, and we highlight the lack of studies that describe its actors from the psycho-sociological point of view.

Keywords Cybercrime · Cyberterrorism · Cyberwar · Psychology

1 Introduction

In the global technological societies, the cybercrime phenomenon is one that keeps on growing. Our society undoubtedly depends on the information and communication technologies (ICT), which also require certain security enhancements in order to be deployed correctly. Privacy, in personal, government communications, banking and ecommerce, is directly related to security, and is at the mercy of potential cybercriminals.

The phenomenon continues to grow, both in terms of individuals, companies, organizations and even governments. Throughout the historical series 2012–2015, there has been an increase in delinquency within the concept of cybercrime. In particular, during the year 2015, a total of 60,154 incidents have been known in Spain, of which 67.9% correspond to computer frauds (scams), that is, 40,846 known computer frauds (Spanish Ministry of the Interior 2015). As an example, according to the 2014 malware data: in that year were created more than 317 million

J.C. Fernández-Rodríguez (✉)
Nebrija University, Madrid, Spain
e-mail: jfernanr@nebrija.es

F. Miralles-Muñoz (✉)
San Pablo CEU University, Madrid, Spain
e-mail: f.miralles@ceu.es

J.M. Ramírez and L.A. García-Segura (eds.), *Cyberspace*, Advanced Sciences and Technologies for Security Applications, DOI 10.1007/978-3-319-54975-0_11

new malicious codes (viruses, worms, Trojans, etc.), which means almost one million threats where launched into cyberspace each day (Symantec Corporation, 2015, quoted by Guilabert 2016).

The case of the engineer John Draper, who was arrested in 1972 for fraud against telephone companies, is often referred as the first cybercrime. In the early 1970s, a famous brand of cereals from the Quaker Oaks company, began to carry out a promotion to increase its sales. This campaign consisted in including a small toy inside each cereal box. These cereals, marketed in the USA, were known as "Cap'n Crunch", and the toy was a small and harmless blue whistle.

John Draper received a call from a blind friend (Joe Engressia), who warned him that, by plugging one of the holes in the whistle, a pure tone of 2600 Hz was produced. A frequency that coincided with the tone emitted by the telephone system, in order to indicate, according to some sources, the end of a call and, therefore, its tariff, and according to other sources, a redirection to another number, which was observed that it allowed to make long distance calls to a free number. This was due to the fact that, in the 1970s, the long-distance voice and data lines used by AT&T (American Telephone and Telegraph Corporation) shared a channel due to the different cost reductions. Once the 2600 Hz tone was emitted, one end of the line was disconnected and the connected side entered in operator mode, ready to "hear" the special tones that determined the call. Literally, this opened the company circuits to anyone who knew how to use them. John Draper, thanks to this idea, devised an apparatus, which was capable of reproducing different tones recognized by the control unit, which allowed him to modify and control the behavior of this for his own benefit. The creation of this artefact, known as "Blue Box" 35 or blue box, became the first great case of what we now know as phreaking or telephone hacking history; its existence soon circled the world with the consequent fame of the inventor. Thanks to Draper, thousands of users were able to make long-distance telephone calls for free, thus causing huge losses to telephone companies, and contributing significantly to the growth of the Hacker movement.

This fact caused a furor among the Electronic Engineering Universities and reached the ears of another young talent of the electronics, called Steve Wozniak, who began to make the already famous boxes and to sell them to make money with his great friend and partner, Cofounder of the company, whose logo is today a bitten apple, Apple. The name of his friend was none other than Steve Jobs.

One of the greatest examples of cybercrime is the recent case of the massive cyber-banking robbery that took place in February 2016, where 81 million dollars were taken from the central bank of Bangladesh and deposited with the Federal Reserve Bank of New York City. Such theft was done by malware that simulated legitimate money transfers through an application called SWIFT, that all banks use to perform transactions between them (Leetaru 2016).

It is common for media outlets, specialized or not, to use expressions with different meanings. This fact can alter the meaning of the report or the news, or even the importance that the readers can give according to their own concept of each term. For this reason, and in order to clarify concepts, it is necessary to provide some definitions that will serve as a basis for future exposure.

The terms cybercrime and cyberdeliquency are used fairly regularly with great similarity. According to Gómez Oliva (2014), cybercrime is often referred to those criminal or illegal activities carried out due to the current media and information and communication systems. If we take into account the definition of the Spanish Royal Academy dictionary, the word "crime" has three meanings:

1. Guilt, breach of the law.
2. Reprehensible action or thing.
3. Voluntary or reckless action or omission punishable by law.

Any of these meanings can be considered as information and communication technologies. Similarly, delinquency is a "set of offenses", and, therefore, the term "cyberdeliquency" can be used in appropriate settings.

The word "cybercrime" is also commonly used. According to the Spanish Royal Academy dictionary, crime is the "voluntary action of killing or seriously injuring someone". Therefore, it does not seem to be a concept that can be applied to the world of communications, although the first meaning of the dictionary can be used, since crime is a "serious crime"; it is also an "improper or reprehensible action", and this meaning can be applied to the virtual world of communications and information technology.

In accordance with the above, it seems logical that the terms cyberdeliquency and cybercrime (the latter understood as a general term, referring to all criminal actions in cyberspace) can be used interchangeably; in the same way, the expressions cybercrime (as a concrete action: "a cybercrime has been committed") and cyberdeliquency can be considered synonymous.

From a more psychosocial perspective, terrorism represents one of the expressions of the materialization of violence, in order to cause terror in the civilian population (Fernández-Rodríguez and Miralles 2016). This form of violence, transferred to the cyberworld leads to what is known as cyberterrorism, which can be considered a form of cybercrime. With the term cyber-terrorists, we mean all those who, whether for one or other reason, attack the integrity, well-being or image of a country (or even a certain territory) or its inhabitants. For example, from the terrorist and traditional terrorist organizations, whose main argument is the religion and the terrorist that use the network as one of their weapons, to the new groups or organized groups of cybercriminals, whose aim is the political or ideological protest, where its members are known as Hacktivists (Mateos 2013).

2 Cybercrime and Cybercriminals

The so-called cyberdeliquency has spread due to the expansion and development of new technologies, gripping a previously unknown space, uncontrollable and with almost unlimited security. A virtual space, which can only be fought with equal weapons which take advantage of new possibilities for the so-called Information Society in the prevention, investigation, testing and repression of the criminal act, overcoming the obstacles presented by the borders of different countries. However, the great challenge of criminal justice in democracies is to achieve, once again, the highest possible level of security and efficiency, with the minimum reduction of guarantees and fundamental freedoms of citizens, within the difficulty, in which criminal law operates, highly sensitive to political and social events, and that at the present moment, especially since the attacks of September 11th, 2001 in New York, it is dangerously inclined to serious sacrifice of guarantees (Francés 2005).

To compile a map of cybercrime in Spain, we must undoubtedly go to the report on Cybercrime in Spain for 2015, published by the Spanish Ministry of the Interior (Spanish Ministry of the Interior 2015). This report has taken into account the statistical information recorded by the Spanish Security Forces (National Police, Civil Guard, Naval Police and different Local Police Corps), which appears in The Statistical Crime System (SEC), as well as from the National Centre for the Protection of Critical Infrastructures (CNPIC).

Besides statistical information broken down into different sections (facts known by criminal categories, number of detentions/imputations, territorial distribution of cybercrime, victim and responsible profile, incidents recorded by CERTSI, Community of reference and by strategic sector, etc.), This report includes an analysis of the information society.

Therefore, along with the statistical data collected in relation to this phenomenon of cybercrime, we introduce data that indicate the use of technologies by the Spanish society in general and in comparison with other countries in our environment. All this is based on surveys and opinion polls conducted by different agencies, not only at national (INE) but also at European level (EUROSTAT).

With more outstanding data, we can highlight the following (Spanish Ministry of the Interior 2015):

- 75.98% of the total number of detentions and accusations made by the Security Forces (5445), correspond to men, mainly, due to the computer frauds, threats and coercion.
- The crimes related to scams, threats, child pornography and the discovery and disclosure of secrets have a higher incidence among the men, according to the analysis of the relationship between the different types of crime covered by this report (Point 4.12), where persons have been arrested/imputed. In addition, among those attributed to women predominate scams, threats, insults and usurpation of marital status.
- The majority of those detained/charged with cybercrime are of Spanish nationality (85.7%) (Point 4.13). Among the detained/imputed of foreign nationality, are those from Romania, Morocco, Colombia, and Nigeria, who bring together the largest number of cases. The detained/imputed subjects are between 26 and 40 years and are frequently involved in the crimes of computer fraud, threats and coercion (Point 4.14). However, regarding to minors, there is a high prevalence in crimes of threats, coercion, access and illegal interception, as well as sexual offenses.

- In the other age groups, the computer frauds, threats and coercions are the most significant in quantitative terms. It should be noted that 28.3% of the accesses and illicit interception of the crimes were committed by minors. Finally, the data on the age of the arrested and imputed persons are listed (Point 4.15). In 2015, the 5445 arrested/imputed recorded in the Spanish SEC, (2397) where in the age between 26 and 40 years.

According to Mateos (2013), there are numerous studies that agree that the typical profile of cybercriminals is a male between 25 and 35 years, and with a minimum computer and technology knowledge, allowing this author to conclude that the Internet is the perfect medium to develop their activities. A cybercrime or hacker, as a rule, distrusts the oppressive authority, and considers that the access to any information that may serve to know the functioning of the world should be unlimited and, of course, free.

In order to understand better the different types of hacker, and not to fall in the mistake of tying them all into the same role, the following profiles will be listed below in a very common classification, used even within the collective itself, and based on very defined patterns of behavior (Mateos 2013).

Black Hat Hacker: They are those who are routinely named as the usual Hacker. They are identified for not following any form of community ethics, and for frequently seeking personal or economic benefit. The Black Hat Hacker looks for ways to collapse servers, enter restricted areas or take control of systems and networks. There may be exceptions, for example, a Russian user of a major hacking forum expressed his discomfort and showed his dislike for hackers who committed attacks on hospitals. In Russian clandestine circles, there is a certain "ethical code" that leaves hospitals out of attack, even though they are in countries that are often targeted by their campaigns and cyber-attacks (McAfee 2016). These people are proud to demonstrate their abilities and their degree of self-realization, which increases when the impact of the injury is greater.

White Hat Hacker: Also known as the traditional Hacker or ethical Hacker. They leave a business card, in order to inform the system administrator about the vulnerabilities or failures found after a foray into its system, and/or performing, in the worst case, as the only modifications, those strictly necessary to maintain its anonymity. Sometimes, White Hat Hacker are subjects that have been part of the Black Hat Hacker, and that they have decided to change their malicious intentions by the support to the administrators of the security systems and the fight against the Cybercrime, using their own knowledge to fight these. The terms Black Hat and White Hat come from the old Western movies, where the good guys wore white hats and the bad guys always wore the black hats.

Grey Hat Hacker: They are subjects whose ethics are of an ambiguous character. These people have similar knowledge to those of Black Hat Hacker, but they use it to find certain vulnerabilities or security failures, which later they offer to solve for an economic payment.

Cracker: They could be included in the Black Hat Hacker group. They are considered the most aggressive group and, perhaps, the most dangerous. Its only objective is, using the expression commonly used by this group, to "burst systems", whether computer or electronic. Crackers are expert programmers, who use their knowledge to modify the behavior of the systems and networks, exploiting any type of vulnerability. They act in an obsessive and insatiable way, guided by their destructive and egotistical eagerness.

Phreaker: Collective focused mainly on the world of telephone systems, including also Mobile Telephone and Voice over IP (VoIP). They know very well the operation of such technologies, as well as their communication protocols. They are dedicated to alter the behavior of such systems, sometimes for pleasure and sometimes for economic purposes.

Lammer: Repudiated within the collective hacker. They are those Internet users who are engaged in gathering information and executing malicious codes, seeking social recognition as hackers without having a real knowledge of the impact of their actions, nor of the operation of the code executed. Sometimes they are really annoying, although their actions do not usually cause great damages.

Scriptkiddie: They are simple internet users with a fondness for the themes of hacking, but without too much knowledge about it. They usually use programs or malware that they find on the network that they execute without more study or knowledge, infecting their own systems in many occasions.

Newbie: Known as hacker learners. They are those novice users who begin to read and experiment with the information found and that sometimes make incursions in weak systems, but without greater significance given their scarce knowledge in the area. Usually their only goal is to learn.

Wannaber: They are those who "want to be" a Hacker with little perseverance and little technical capacity; the great majority are harmless. They often use their limited knowledge to gain social recognition outside the network.

Informatics pirates: Although this term is often confused with Hacker, the informatics pirates are only engaged in the copying and distribution of software, music, games and contents in an illegal way, attacking the intellectual property and the rights of its owners.

Buccaneers: They play the role of merchants on the network. They buy and sell illegal material obtained through others, such as identities, access control cards, cracked software, etc.

Continuing with the terminology used, it seems safe to consider as cybercriminal anyone accused of carrying out cyberdeliquency or cybercrime acts. Thus, they are individuals whose illicit activities, previously pursued, have evolved with technology and have expanded their horizons in the network, such as pederasts, pimps, etc.

Although we have mentioned the different types of hacker, or subjects dedicated to cybercrime according to their way of acting, it is very interesting to add some additional profiles that have appeared today and are based on different technological articles. Highlighting, among them, the Report on Virtual Criminology that the computer security company McAfee performs each year (McAfee 2009, quoted by Mateos 2013; McAfee 2016).

Bots fitters: They are those users whose intention is to gain control of a remote computer through the installation of malicious software. To achieve these purposes, they use pre-programmed malware, which is hidden in all kinds of interactions that the user makes while surfing the web.

Carders: These types of cybercriminals focus exclusively on identifying the theft and credit card fraud on the Web. The carders can be seen as the natural evolution of the traditional street pickpockets. Once they have obtained the necessary information, they can carry out online transactions and purchases by covering up their identity and charging the cost to their victim.

Cyberpunks: Without having a lucrative objective in their actions, the cyberpunks, which comes from a literary movement with the same name, can produce great losses to its victims, such as economic losses and image losses. Considered as the mischievous cybercrime, cyberpunk is dedicated to altering public systems, such as a web page, in order to mock and ridicule those users, who they consider their victims.

Insiders: They are employees or former employees that act within the companies, in which they work or have worked, using their experience and knowledge of the "inside" systems, in order to access and distribute confidential information or to cause damages in their companies. Their motivations are usually both economic and personal, even for revenge purposes.

Phisher, Spammer: Users who are specialized in using e-mail as a way of communication with their victims. They try to achieve an economic benefit through deception and lure that lead to confusion, since they are shown as seemingly reliable sources.

The individual who commits these types of cybercrimes is not considered a common criminal, since it differs in the mechanism and the means used, in order to produce the result. It is important to mention, that there is no individual profile that can describe these criminals, extracting only a series of common characteristics (Garrido et al. 2006). We can affirm that the people who commit these criminal behaviors have certain characteristics that the traditional criminals do not have: they have high abilities in the handling of computer systems and, normally, they have sensitive information in their workstations (Gallego 2012).

An investigation by the Yale University of the United States, which studied the characteristics of cybercriminals, shows that these individuals who commit computer fraud, usually have the following characteristics: most are middle-aged, married, and economically stable; they have a fixed job, a high level of education, good self-esteem and do not consider themselves delinquents (Garrido et al. 2006, quoted by Dinca 2016).

Criminological studies that have traditionally dealt with the profiles of these criminals describe subjects who are motivated to commit crimes without the need to have specific knowledge. The subjects that carry out these criminal practices must have a minimum computing knowledge; without that knowledge they will not be able to participate in the criminal acts. Thus, although it cannot really be said that this type of criminals does not focus on a particular social class, if we can affirm that it is illegal, it is restricted to a particular group of criminals, at least those who have a certain level of intelligence or character. This intellectual level allows them to have enough computer skills in order to access the different systems (La Cuesta and Pérez Machío 2010).

Another feature that makes the new technologies attractive for cybercriminals, especially for cyber-attacks of different types, is the massive effect that their actions can have. With the Trojan infection of hundreds of thousands of computers, it is feasible to launch simultaneously attacks with really serious consequences. Thus, the most daunting task following these cases is the subsequent investigation of the origin of these incidents, because it is very difficult to get to the real cause of the damage: the attacks come from thousands of infected computers from different states, whose owners may even not be aware that they have been part of the infrastructure used to carry out the attacks. This fact presents a great difficulty to identify the real author of the attacks and, therefore, these authors can achieve an attractive anonymity. This anonymity can also come from other circumstances than the ones mentioned in this article, since there are techniques that allow the camouflage and, to some extent, to hide the direction of some equipment (Gómez Oliva 2014).

A particular type of cybercrime involves the cybercriminal who operates within the companies. They may not be great technology experts, but they know the security vulnerabilities of an organization and take advantage of them, in order to extract the data they are looking for. This type of criminals are called insiders, whose simplest case may be the executive, who changes work and takes the USB database of the clients with he has been working with (Portafolio 2013).

According to the mentioned publication, the insiders generally do not need to be great computer experts; they just need the knowledge and the criterion of knowing how to distinguish between the correct and incorrect information that interests them, and the way, in which they will obtain it. Thus, we can mention some risk indicators that must be taken into account, in order to detect a cybercrime within a company:

- People with a good knowledge of the user and the network. They constantly use computers without explaining what they do and without sharing that knowledge, even despising others for their computer skills.
- Employees who work more for no apparent reason. They do not enjoy holidays, or do not make frequent use of computers in work, but at home.
- A very serious warning situation is when people whose work tasks have nothing to do with computers, such as cleaners, are using them in the office.

Digiware, an integrative security company, completes these points with several additional features (El Heraldo 2016):

- People who take advantage of social spaces to be interested in customer data and other information of restricted or particular use.
- People who install spy programs without authorization from the organization.

- People who deactivate antivirus software on their computer.
- People who use the computers or devices of other members of the organization without authorization.

However, there are also inexperienced people who know the vulnerabilities in the security of the company and they take advantage to extract the data they need.

Following Digiware's study, nowadays, cybercriminals don't act alone, but operating as a part of large criminal organizations around the world, attacking roughly 600,000 times per day, mostly the financial and government sectors.

Digiware claims that 50% of cybercrime groups are usually made up of 6 or more people. Among its components, 76% are men, whose ages are between 14 years (8%) and 50 (11%). Although the average age of this type of criminal is 35 years (43%). 50% of cybercriminal groups have operated for more than six months, while 25% have operated for 6 months or less. Most of their activity is registered in North America and South America, with 19% of the total attacks generated worldwide.

Digiware also claims that most attacks inside companies are carried out by a hacker in a premeditated way, and with direct intervention in the internal systems of the organizations. These accesses are due to negligence of the employees (9.3%) and due to accidents of the members of the company (7.3%).

Digiware notes that the trademarks attacks are the most commonly used by criminals, in order to reduce the stock value or to seriously affect the company image through frauds to their customers. Furthermore, they try to interrupt digital services, such as blocking access to emails or hacking websites. Taking into account that all these "services" have different rates, such as 25 USD$ for stealing a Skype account or 200 USD$ for accessing data on social networks and accessing professional data.

In reality, any person with enough knowledge of computer science, can lend his work to security companies or use it for his own benefit. It is also of concern that the profile of these Black hats responds more and more frequently to young people, in many cases underage, which in the latter case still damages the counterattack or the response. Regarding this fact, it is worth noting that the criminal responsibility of minors (at least in the case of Spain) is required for people over the age of fourteen and under the age of eighteen for the commission of acts classified as crimes in the Penal Code or in the Special criminal laws. It is called a black hat hacker, who penetrates the security of systems to obtain a personal gain or simply by malice; it is called a white hat hacker, who penetrates security systems to find vulnerabilities (de Dios 2013).

Taking all this into account, it is necessary to warn about the impossibility of creating a single profile of cybercrime. This issue is explained by the intimate relationship that the cybercriminal shares with cybercrime, adding also the difficulty of the judicial persecution of this type of delinquent. On the other hand, it is necessary to emphasize that the absence of quantitative and qualitative criminological studies on concrete profiles does not offer a single general conclusion (Miró 2012).

For example, in the case of the economic cybercriminal, the profile does not respond to a concrete and individual subject, but to a criminal organization, being the paradigm of this cybercrime the hacker. In political cybercrime, it emphasizes the different organizations that it presents in relation to the criminal agents of the physical world, with emphasis on its horizontal and non-vertical hierarchy, with a central objective, to which are united multiple people, who share ideologies without the need of high technical knowledge. Equally, the possible individual performance, outside of any organizational relationship, is not ruled out. Finally, the social cybercrime, which may be the most complex, especially by the existence of multiple motivations in the active subject, but always with the particularity that permeates cyberspace (di Piero 2012).

3 Cyberterrorism and Cyberterrorist

Cyberterrorism goes beyond cybercrime, although many scholars consider that both issues are the same. Certainly, they have some connection, because on multiple occasions, cyberterrorists carry out criminal activities on the network, but the causes that motivate their actions and the benefits they expect to receive from each other are different. Cyberterrorism is the convergence of cyberspace and terrorism, that is, the way in which terrorism uses information technology to intimidate, coerce or cause damage to social groups for political-religious purposes. It is the evolution that results from changing weapons, bombs and missiles for a computer, in order to plan and execute attacks that produce the most serious damages possible to the civilian population. This question, therefore, implies a great difference in relation to cybercrime. Cyberterrorism seeks to cause the greatest possible harm for reasons that are usually politico-religious, while cybercrime actions are aimed at obtaining a profit mainly of economic type (Sánchez Medero 2012).

How can we get to know the cyberterrorist? Due to the technical knowledge, we can assume that he is a person with a high intellectual coefficient, but according to the new times, and with the new generations, we could not consider it a preponderant pattern. The cyberterrorist can act without the need to have an excellent profile in the handling of computer science. According to Patron (2013), some of the characteristics of a computer terrorist are:

- It is a person who through his actions causes panic and terror with the purpose of weakening and discrediting governments, society, a belief, etc.
- He is a person with a strategic character, because he is able to capture both the attention and the support of the community (colleges, universities, churches, etc.). In this case, he uses computer media and social networks.
- It is a person who has clearly marked his targets; both target attack and target population (target audience).
- His criminal profile is far from the parameter of "white-collar criminal", as the American sociologist Sutherland (1943) pointed out earlier to refer to a

computer criminal. With the advancement of the times, to think that only people with high purchasing power can be cybercriminals, is to make a clear mistake in the world perspective and its progress. Anyone can be trained to be a computer terrorist, just as anyone can be trained to design web pages; you only need a computer and possibilities to learn.

- A person with a strong resentment towards society or a group. This is, perhaps, a point of the emergence of terrorism.
- Finally, the person knows the ways of attacking by using different computer media. This point is understood as knowing what programs to handle, how to get people through the computer media and how to reach their various purposes with the use of the network.

4 Cyberwar

First, we must make a difference between cyberwar and cybercrime. The objective of the latter is the economic retribution or the fact of causing damages to the victims, leaving aside the activist ideology. Cybercrime uses tools and methods similar to the cyberwar, but their goals are different (Pantano 2014).

Cyberwar can be defined as a type of aggression promoted by a state and that seeks to seriously damage the capabilities of another state to impose the acceptance of a goal, or simply, in order to have information, alter or destroy their communication systems, to modify their databases. The main difference with what we have traditionally understood as war, is that the medium used would not be the physical violence, but a computer attack that goes from "infiltration in enemy computer systems, in order to obtain Information to the control of projectiles through computers, through the planning of operations, management of supply, etc." (Colle 2000). Cyberwar involves the use of all electronic and computer tools to crush the enemy's electronic and communication systems, and to keep their own operational resources (Sánchez 2008).

Due to special characteristics of this phenomenon, no data has been found on the profiles of its different actors.

5 Conclusion

Network through internet has become a perfect space for the implementation of cybercrime and cyberterrorism. It is a mass medium that offers easy access, little or no governmental control, and anonymity, as well as rapid flow of information, high impact, and low risk, cheap and undetectable in most cases. In addition, in spite of the work done by the agencies or gubernamental secretaries of security, it is very complicated to guarantee the complete security of the computer systems.

Although there is always the option of not connecting to the internet, in our days this option is not logical or reasonable, despite the exceptions imposed by some countries, such as China, North Korea or Saudi Arabia. We must remember that China and North Korea have often been accused of cyberattacks.

There is also another possibility of identifying vulnerabilities and identifying the existing and potential dangers that such weaknesses allow, and try to strengthen computer systems. An increased control of communications, the creation of specific agencies and, therefore, cyber-guards, at the moment, in our opinion, do not seem to have positive results.

So far, and according to the revised documentation, there are no specific profiles that allow us to identify cybercriminals and cyber-terrorists for diagnostic purposes; we do not have such information on cyberwar actors. Therefore, the data available do not go beyond a simple description of general characteristics. It is true that the data obtained are helping to detect cybercriminals and cyber-terrorists, but they are not enough to control or prevent the activity of these people in the network.

In any case, as described throughout this document, cybercrime, cyberterrorism and the governments are developing the network in order to exercise, extend and develop their activities, although, with different goals and objectives.

Cybercriminals use internet for purposes such as defrauding, damaging and blocking, in order to achieve economic profitability or reaching their interests outside the law. Terrorist groups are shifting their diffuse organizations into cyberspace, as a way to dilute themselves in a place that is complicated to counteract, and in this way, they are using the network to fund, recruit, train, communicate, coordinate, indoctrinate, advertise, etc. to continue to maintain their organizations and achieve their objectives.

Some states have succeeded in transforming cyberspace into a new battlefield, which gives them more advantages than the traditional one. Since cyberwar is characterized by its great asymmetry, short duration, rapid reaction, low economic costs, but, specially, to cause minor physical damages to the soldiers. Also, it is a greater space for combat, including the possibility of intense fighting due to the superiority of the information, alongside an increase of the integration, including attacks that can be launched from any place and almost be undetectable.

In conclusion, and contrary to in other forms of confrontation, in cyberwar it is not necessary to have sophisticated weapons and a large and numerous army or to be close to the "target". All one has to have is certain computer skills. In addition, this type of confrontation can begin from any part of the world, and even, simultaneously, start from distant places, without having to assume great risks. As if this question were not enough, the continued appearance of new computer tools, as well as their free access and propagation, makes the identification of the alleged attacker more complicated and easier to maintain the anonymity. Therefore, its effects can be equally devastating than the traditional war. In this way, there has been a great expansion of the range of actors that can intervene and create a conflict. However, both states and terrorist groups are making passive use of the network, with the exception of the Islamic State and related groups.[1]

On the other hand, we cannot say the same thing about cybercriminals. Although in any case, we believe that, sooner or later, both will make more active use of cyberspace to perpetuate and carry out their actions. But we are not the only ones, who think of this fact. In the annual report of the security company McAfee, it is argued that we are on the way to a "cybernetic cold war". Hence, we can say that cyberwar, cybercrime and cyberterrorism are one of the biggest threats that we will have to face in the 21st century.

As users of the network, we also have the responsibility to try to get our information to the chosen destination and to use it correctly, or for the original purpose. While we use the internet, we must have good practices: the best antivirus and keep the software we use up to date. In one way or another, with the means at our disposal, we must try to ensure that everything we send on the internet have a good reputation and cannot be intercepted.

Nowadays, cybercrime is configured as a new vision of what we could call a normal crime. In a globalized world, where the population is 24 h a day connected to the network, criminals have found a new scenario, full of possibilities, where they wait for the right moment to execute their crimes.

As with traditional crime, cybercriminals are not a person with a single profile either, but it is possible to distinguish many types according to one or another methodology, according to their personal objectives or their role in a criminal organized structure.

Organizations, companies, governments, and citizens are exposed to a whole catalogue of crimes and scams in the network, although they don't always play the role of victim. In some cases, the perception of anonymity and freedom that can be breathed on the Internet, leads to "neglect" the attention on where the border between legal and illegal lies.

References

Colle R (2000) Internet: un cuerpo enfermo y un campo de batalla. Revista Latina de Comunicación Social, 30. Retrevied from http://www.revistalatinacs.org/aa2000qjn/91colle.htm

de Dios J (2013) Los nuevos modi operandi de los Ciberdelincuentes durante la crisis económica. Revista de Derecho UNED 12:495–523

de La Cuesta, JL, Pérez Machío, AI (2010) Ciberdelincuentes y cibervíctimas. In: de La Cuesta JL, de la Mata NJ (ed) Derecho Penal Informático, Civitas Ediciones. Thomson Reuters, Navarra, pp 99–120

di Piero C (2012) El cibercrimen. Fenomenología y criminología de la delincuencia en el ciberespacio. Retrevied from http://www.indret.com/pdf/984.pdf

[1]See chapter … by Claudio Payá.

Dinca CF (2016) Fraudes en internet. Trabajo final de grado, Universidad Jaime I. Retrevied from http://repositori.uji.es/xmlui/bitstream/handle/10234/161252/TFG_2016_DincaClaudia.pdf?sequence=1
El Heraldo (2016) Conozca el perfil de un ciberdelincuente, según Digiware. Retrevied from http://www.elheraldo.co/tecnologia/conozca-el-perfil-del-ciberdelincuente-258538
Fernández-Rodríguez JC, Miralles F (2016) The terrorist suicide woman in Jihadism. In: Ramírez JM, Fernández-Rodríguez JC (eds) Security in infraestructures. Cambridge Scholars Publishing, Newcastle, pp 186–202
Francés MLG (2005) Reflexiones sobre la ciberdelincuencia hoy (en torno a la Ley penal en el espacio virtual). REDUR (Revista electrónica del Departamento de Derecho de la Universidad de La Rioja) (3):4
Gallego A (2012) Delitos informáticos: malware, fraudes y estafas a través de la red y cómo prevenirlos. Retrevied from http://e-archivo.uc3m.es/bitstream/handle/10016/16868/pfc_alberto_gallego_yuste.pdf?sequence=1
Garrido V, Stangeland P, Redondo S (2006) Principios de criminología. Tirant lo Blanch, Valencia
Gómez Oliva A (2014) Cibercrimen y ciberterrorismo. ¿Exageración mediática o realidad?. Trabajo de fin de grado. Universidad Politécnica de Madrid
Guilabert NG (2016) Actividades cotidianas de los jóvenes en internet y victimización por malware. Revista de internet, derecho y política 22:59–72
Leetaru K (2016) What the Bangladesh SWIFT hack teaches about the future of cybersecurity and cyberwar. Retrevied from http://www.forbes.com/sites/kalevleetaru/2016/04/30/what-the-bangladesh-swift-hack-teaches-about-the-future-of-cybersecurity-and-cyberwar/
Mateos I (2013) Ciberdelincuencia Desarrollo y persecución tecnológica. Retrevied from file:///C:/Users/jcfrada/Downloads/pag%2020%20a%20la%2026%20perfil%20ciberdelincuente.pdf
McAfee (2016) Informe de McAfee labs sobre amenazas Sept 2016. Retrevied from http://www.mcafee.com/es/resources/reports/rp-quarterly-threats-sep-2016.pdf
Miró F (2012) El cibercrimen. Fenomenología y criminología de la delincuencia en el ciberespacio. Marcial Pons, Madrid
Pantano A (2014) Ciberguerra. Retrevied from https://dspace.palermo.edu:8443/dspace/bitstream/handle/10226/1448/Ciberguerra-Pantano%2068586.pdf?sequence=1&isAllowed=y
Patron P (2013) Drogas informáticas Y terrorismo Informático ¿Nuevos métodos de delito o mitos digitales?. Retrevied from http://www.derecho.usmp.edu.pe/cedetec/articulos.html
Portafolio (2013) Así es el perfil del ciberdelincuente en las empresas. Retrevied from http://www.portafolio.co/tendencias/perfil-ciberdelincuente-empresas-67906
Sánchez G (2008) Ciberterrorismo: La guerra del siglo XXI. El Viejo Topo 242:15–24
Sánchez Medero G (2012) Cibercrimen, ciberterrorismo y ciberguerra: los nuevos desafíos del s. XXI. Revista CENIPEC, 31. Retrevied from http://www.saber.ula.ve/handle/123456789/36770
Spanish Ministry of the Interior (2015) Estudio sobre la cibercriminalidad en España. Retrieved from http://www.interior.gob.es/documents/10180/3066430/Informe+CibercrimiCibercr+2015.pdf/c10f398a-8552-430c-9b7f-81d9cc8e751b
Sutherland E (1943) Juvenile delinquency and urban areas: a study of rates of delinquents in relation to differential characteristics of local communities in american cities. Am J Sociol 49:100–101

Author Biographies

Juan Carlos Fernández-Rodríguez is a Professor at Universidad Antonio de Nebrija of Madrid, being Director of Postgraduate Studies in Risk Prevention, Director of Academic Degrees at the School of Social Sciences, and a member of the Scientific Council and its Chair on Risk and Conflicts Managment. Dr. Fernández-Rodriguez has advanced degrees in Psychology—

Licenciatura from Oviedo University and Ph.D. from Universidad Complutense of Madrid, with the unanimous outstanding rating *suma cum laude*—Technician in Occupational Risk Prevention, and a University Expert in the Management of People in Global Environments. He has participated in different research projects dealing with Educational Technologies, the Knowledge Economy, and Globalization, Managing stress and other psychological topics. He has written several papers in scientific journals and books and contributed to both national and international conferences, not only with scientific communications but also as a member of their organizing committees, such as the XLI CICA, from which this book is a scientific product. Dr. Fernández-Rodriguez also has extensive experience in academic and research management and evaluation, as well as in teaching innovation projects and curricula.

Fernando Miralles-Muñoz is a Professor at Universidad San Pablo CEU of Madrid. Dr. Miralles has advanced degrees in Psychology—Licenciatura and Ph.D. from Universidad Complutense of Madrid, Official Clinical Psychologist. Previously he was an Associate professor at Carlos III University, Universidad Complutense of Madrid and Universidad Pontificia de Comillas. He has participated in different research projects dealing with Anxiety, Personality, Communication and various mental disorders in the Armed Forces in Spain. He has extensive teaching and university management experience.

Use of Cyberspace for Terrorist Purposes

Claudio Augusto Payá-Santos and Juan José Delgado-Morán

Abstract New technologies have created a new battlefield, with new targets to fight. In cyberspace, the answers of local and international authorities, have been different, being specially important the counterterrorist policies, as well as the scanning and infiltration of intelligence services, of activities and communications, in order to prevent terrorist acts and get physical evidence to be used in front of a court, by creating special units, like the Spanish Command of Cyberdefense. In light of this, different policies have been implemented focused on cybersecurity, like the creation of specialized centers like the European Cybercrime Center or the US Threat Intelligence Integration Center. This chapter offers some analytics about the use that ISIS is giving to cyberspace, from recruitment purposes, to propaganda, financial support or even psycho-war. We will also analyze the tools used for these purposes, like the Deep Web, Social networking, including Facebook, Instagram or Twitter, as well as other social networks created by themselves or their own media group.

Keywords Cyberspace · Terrorism · Cybersecurity · Cyber defense

1 Introduction

There are many theories about the same reality. Kaldor (2001) says that during the 80s and 90s new wars started to appear with targets linked to identity policies. These new conflicts of our Global Era, born in an environment characterized by the

Submitted: 15.10.16; Accepted: 6.12.16.

C.A. Payá-Santos (✉) · J.J. Delgado-Morán
Security and Defense Studies, Nebrija University, Madrid, Spain
e-mail: cpaya@nebrija.es

J.J. Delgado-Morán
e-mail: jdelgado@nebrija.es

J.M. Ramírez and L.A. García-Segura (eds.), *Cyberspace*, Advanced Sciences and Technologies for Security Applications, DOI 10.1007/978-3-319-54975-0_12

absence of certain structures of State, as well as, the impossibility to exercise the monopoly of the legal use of military power. According to the aforementioned author:

> The goal is the control of the population, removing anybody who has a different identity. So, the strategical target of these wars is to pull out the population through several methods like massacres, forced relocations and many more politic, psychologic and economic intimidation tactics (Kaldor 2001, p. 4).

In a similar way, the traditional differences between civil and military, between fighters and non-fighters, have disappeared in this new kind of conflict; for example, the participation of paramilitary groups, asking for money to supply protection (Kaplan 1994). Another feature of these new conflicts are the innovations in weapons and tools used in them. Nowadays, in a world influenced by internet use, the destructive capability of first and third world countries is much closer, making it possible to cause important damages to the enemy without needing big military power.

Reflecting on this point, it is very important to insist in the possibility of understanding the global terrorism as a new player in this new kind of conflicts featuring a common characteristic: asymmetry. The US Army knows asymmetry in military terms: "to act, think and organize in a different way to the enemy, in order to maximize our strengths and take benefit of the weak points of him" (Patterson 2002).

This asymmetry is based on the enormous vulnerability of Western Societies against Jihadist terrorism. Many analysts think that the main reason of success of the Islamic State of Iraq and Syria (ISIS) is their renovated skills. In order to build a caliphate, the leaders of ISIS or Daesh (Daesh or Da'ish is an acronym for al-Dawla al-Islamiya fi al-Iraq wa al-Sham, the real name of the Islamic State, and was used by the organization itself in the beginning).

Daesh has created a strategy of propaganda among terrorist groups. Their communication strategy tries to copy the institutional communication of a state. Their innovation is to start using media allocated on cyberspace, like social networking, Twitter, Facebook or Instagram, adapting their communication to their own goals.

Until now, we didn't have a general and globally accepted definition of cyberterrorism. Internet has become a social, cultural, economic and technological phenomenon that approaches people to institutions in an easy, fast and cheap way. Besides that, the net allows us to avoid all physical restrictions and generates a bigger volume of interactions with a global community that doesn't need a geographic location or physical contact between them (Sánchez Medero 2010).

In the 80s, the term was started to be used in definitions like "convergence of cyberspace with terrorism" and, in the 90s like "the cyberterrorism is the premeditated and politically motivated attack against the information, systems, software and data, from terrorist groups or secret agents of foreigner countries". By cyberterrorism or electronic terrorism can be understood the use of technical, information, communication, computer, electronic or similar means, with the purpose of generating terror or fear among a population or government in general, causing a violation of the free will of people This can be done with political, economic or religious purposes.

In recent years, the rise of technology has increased in the political, social and economic spheres. Nowadays, Internet is the biggest social phenomenon and influence people´s lives daily. That phenomenon is allowing access to millions of websites everywhere, with all kind of information, and a fast and global system of communication, but, as usually happens sometimes, this instrument is also been used to satisfy illegal interests of individuals and groups. So, every day more and more cases of illegal acts through the net can be found. For this reason, the net is a productive power, that has given us a new generation of "part time terrorists", that are able to carry out a strong commitment with the jihad, as well as a well as carry out professional activities and a social life apparently normal. Henceforth cyber-activists cooperate with terrorist's groups without the need of being at the original location of that organization (Castells 1999).

As a result, through cyberterrorism, we can see the use of cyber-tools, to stop, harm or deny access to critical national infrastructures, like energy, transport, communications or government services, with the purpose of coercing or intimidating a government or civil population. This is clearly a rising threat and we must develop preventive, deterrent and countering skills in order to stop this menace.

2 Is the Use of Cyberspace a Cyber-Threat?

Communication through the Internet is cheap, effective and simple. You just need a PC or a mobile device with internet connection to access all kinds of contents. Hence, the use of cyberspace as a conflict scenario is an important topic of discussion right now as cyberterrorism is an attractive opportunity for terrorist organizations.

Considering this definition, we are hesitant about the capabilities of current terrorists in taking control of this new global threat. Obviously, the possibility of making a "mega-attempt" can awake the interest of certain terrorist groups, but these acts are not so easy to perform. Organizations like Al Qaeda have experienced with these means without achieving fruitful results. Their acts have been limited to practices that every "self-made hacker" can do, like websites sabotages, and stealing personal data (Jordán 2013).

According to the former U.S.A. National Director of Intelligence, Mike McConnel, terrorist groups will archive cyber-sophistication, in the same way as atomic proliferation, but easier (Nye Jr 2011). Cyberterrorism, opposite to many opinions, only implies the use of cyberspace as a tool to inflict physical injuries to people or things. This is much more complex than hacktivism, the use of internet for propaganda, financial help, obtaining information or private communications among their members (Torres Soriano 2015).

The appearance of internet, its extended use, and its potential have transformed the "net of nets" in a tool for terrorists giving way to a new scenario. According to Ban-Ki-Moon, the former U.N. General Secretary, "internet is an excellent example of how terrorists can act in a truly transnational way. Thus states must think and work also in a transnational way (The Global Information Technology Report 2015, p. 20)".

Weiman (2004) has identified at least 6 different ways of terrorist use of cyberspace:

1. As an instrument of psychological war, through the diffusion of images causing terror among the enemy population, (like images showing hostages executed by decapitation).
2. As an instrument of propaganda. Terrorist organizations can advertise their actions in live emissions to all the world, once they have made up information, in order to maximize their achievements and minimize their mistakes.
3. As a financial instrument. Al Qaeda got financial help thanks to the personal fortune of Bin Laden as well as the contribution of several non-governmental organizations (NGO). Nowadays, experts like Jimmy Gurule, are pointing to Bitcoin like the suitable channel for financial support to terrorism. Activities like smuggling petrol, led by Daesh, can be cashed through payments with this cyber-currency.
4. As a recruitment instrument. Using the net, Daesh has multiplied the number of foreign fighters that Al Qaeda had some time ago. The massive diffusion among the population, of images and videos showing the fair side of mujahidin's' life, as well as their success against the non-Muslims enemies (including executions), has helped DAESH keep open their information and recruitment offices in the world on a permanent basis. The success of these methods among the young Muslims all over the world has been important, as expected.
5. As a networking and structures occultation instrument. The hierarchic organization of terrorist groups has been practically hidden due to the establishment of structures of communication in the net. Vertical structures have been blurred against horizontal structures, so the members or cells of different terrorist groups, can support each other, co-ordinate and plan attacks in a cheaper and safer way. In the deep net, chats exist where members of several terrorist cells can co-operate in order to plan their attacks in a coordinated way, applying and checking with the management of the organization and so. Al Qaeda recently made a call to all the "jihad brothers" to use PalTalk in order to not be detected.
6. As a repository of document. We can find in the web many handbooks and guides about making explosives, waging urban combat and guerrilla tactics.

Under these circumstances the fight against terrorism on the net, has been focused on tracking communications among terrorists that have attempted or were planning to commit acts as well as the tracking of the people responsible for these organizations. The resulting investigations have proven the dependence among neutralized terrorists, as well as the huge computing capability that some terrorist organizations have right now. In the words of Jaquelyn S. Porth, a File Security Affairs Writer for the US State Department:

> Internet has expanded dramatically the recruitment, training, motivating and co-ordination capability of radical groups, in long ranges without direct contact. Terrorists can check websites to learn how to shoot down helicopters, see hostage's decapitations, read letters written by the kamikazes or listen to the messages of their leaders. Even without websites, Internet allows diffusion of radical messages, as well as operational instructions sent by e-mail (Porth 2007).

3 Cyberspace at the Service of Terror

3.1 Cyberspace at the Service of Daesh

Daesh has become a world reference of the *jihad*, overcoming Al Qaeda in many aspects. Many experts in cybersecurity, such as Michael Rogers, a US Navy Admiral and Director of the U.S. National Security Agency, think that nowadays Daesh doesn't have yet the suitable means to launch a massive attack against the western countries, though the acquisition of those capabilities is a matter of short time. While all the Western countries were shocked at the terrorist attack against the French magazine Charlie Hebdo, members of Daesh got into the website of Malaysia Airlines and blocked it with the following message:

> Error 404-airplane not found. ISIS will win.

Another exhibition of their capabilities, even worse, was the hacking of the Twitter account of one of the workers of USCENTCOM, publishing two messages under their name. As an example, we will mention 2 extracts of each one: first message was: "American soldiers, we are coming, be careful. ISIS". The second one started with "…The Cybercaliphate, supported by ISIS, goes on cyberyihad (Seck 2016).

This attack suffered by the US Military Administration has shown the world not only the existence of jihadists with computer skills, but also that the US Government was not able to neutralize the origin of the attack. It's also relevant to consider that, only during 2013, the economic loss of the 13022 cyberattacks carried out against the USA, China and Germany by criminals exceeded 177.000.000 € (Symantec 2013).

Fortwith the self-proclaimed Islamic Estate is becoming a true weapon in the social networking arena. The group has used all kind of social networking—Facebook, WhatsApp, Instagram, Twitter, Skype—and another application not so well known like KIK or Diaspora, to extend their influence among all the Muslim users as well as to try to recruit new followers. For example, Diaspora is a platform launched in 2010, promoted by a crowdfunding campaign created by 4 students from New York. The support team of Diaspora, has admitted that a big number of members of IS had opened accounts in this platform, after all the restrictions in Twitter, Facebook or YouTube. This has in part, due to the format of this platform, where it is not possible to manipulate the publications, the messages they publish, and cannot be deleted. Facebook is perhaps the social network most used by the terrorists, especially as a way show off brutally of the Al Baghdadi the faithless. Their objective was to share images of decapitations, bags full of arms and legs and, heads piled up on the path. In all the publications, they make fun of an enemy's death in order to achieve stronger radicalization and a later affiliation. Facebook has been used also as a financial tool. Another social network highly used Instagram and its objectives are quite similar to the ones of normal users. Despite the fact that they are not publishing photos in paradise beaches, followers of Daesh upload pictures to show off their lives and make people jealous.

This similitude between the use of Instagram among Western users and jihadists, is due to the fact that the majority of new recruits are coming from first world countries. Thus jihadists try to convince their Western friends that have not decided yet to join them, through photos where luxury, experiences and a life full of adrenaline are the main incentive. Terrorists are also using Alrawi as a tool to organize their attacks (Barrancos Larráyoz 2014).

Alrawi app is an application that can be used in Android devices, but is not available in Google Play and it can be downloaded from different internet websites. Another of the apps involved with the terrorist group, that has been unveiled is Amaq Agency. In this app, they were delivering news and jihadist propaganda. The app is called Amaq News and was announced for Android trough social networks close to ISIS, in particular the News Agency Amaq.

5Elafabook (translated like Califatbook) is a platform created by followers of IS (number 5 or Hamsa, is used in the chats in Arabic to represent the sound of the j). Companies like Twitter have already started to close accounts, closing more than 1000 during the last year, mostly after publications of bloodthirsty videos where journalists or civilians were decapitated. Consequently, more than 46,000 Twitter accounts, have moved to other platforms like JustPaste, SoundCloud, VideoPress or Instagram. Even Google has deleted some accounts due to the popularity of the question "How to join the IS?"

The digital networking efforts of Daesh can be compared with the ones carried out by any State or Western company. Their readiness to create software is evident with the creation of their own original apps like the one called "The Dawn of Glad Tidings", that was available in Google Play Store from Android.

The digital magazine Dabiq has edited more than 10 issues, publishing periodically information about the caliphate. This magazine has English and French covers, and usually includes around 50 colored pages with pictures and text. The subjects of the magazine are focused on all the main points of the IS: *Jamaa* (community), *Hegira* (migration), *Manhaj* (chasing of the truth), *Tawhid* (unity) and, of course, *Jihad* (holy war). The name of the magazine was inspired on a battle in Siria, that certain Muslim myths link with the Apocalypse. The idea of the terrorists is to give an impression of an apocalyptic message, for instance, the front page where the Daesh flag is on the obelisk of Vatican.

From a technical point of view, videos of Daesh don't have anything to envy from the ones created by Hollywood industries. There are many different levels of professionalism, from recordings with smartphones, to much curated productions. It´s obvious that many followers have the skills of digital modelling, photography and video-edition, implying that there are many production companies working directly for the caliphate. One of them is Al-Furqan; another one is Al-Hayat Media Center. According to Barranco Larráyoz (2014), analyst of THIBER, a think tank specialized in cyber-safety, propagandistic videos of Daesh have 4 targets:

1. Create fear among the soldiers of enemy armies
2. Increase the support to the organization
3. Recruit new members
4. Create new links with other terrorist organizations.

Every video that this organization issues with narrative resources is used to persuade the audience, re-shape their minds and influence their thoughts and actions (Orellana 2013). The priority in the message of Al-Hayat Media Center Productions is the call to the Jihad. Hence, this call is announced changing the scenario and showing different faces of the organization, depending on the target of people to who it is addressed. As a result, there are five sorts of videos distributed by Daesh: Nashid, Arenga, Mujatwits, Report and Documentary. Let´s have a closer look at each one:

1. Nashid: this kind of video is composed of several images that go together with a text of a song or Nashid. According to Said Benham, expert in this sort of religious songs, Nashid is a choral music, *a capela* or accompanied by drums. Originally, their texts referred to the history and the faith of Islam, but lately, they have included political subjects. However, the Daesh doesn't allow the use of instruments. These songs refer not only to rational thoughts but also to the soul, and work like an element of seduction for those who practice religion like a link among jihadists, creating a narrative and collective idea. Nashid texts encourage to fight, praise liberty and to recover dignity. The first songs created by Al-Hayat, were dedicated to hate, violence and martyrs. In these clips, there are always soldiers marching, smiling and porting their guns. The war is introduced by Daesh like an adventure, followed by an easy and quick victory that turns the brave fighter into a hero. All the special effects intentionally applied in postproduction works have the goal of creating an expecting and exciting rhythm. Therefore, the Nashid is a kind of song included in a video-clip focused only on followers of the terrorist group (Marshall 2015).
2. Arenga: this kind of video-clip is filmed outdoor, with one or more speakers dressed as fighters, inviting the public to join the Jihad. Here the image is subordinate to the talks. In this category of clips we find images of executions, although they are not the main subject of the production company, where speakers threaten the enemy pointed out and make it responsible of all their disgraces. The camera usually makes close takes of the horrified faces of the victims. Normally they use to slow the speed of filming in the moment of the

execution, increasing, in that way, the terror among the audience (De la Fuente 2016).

3. Mujatwits: a hybrid word from mujahedeen and twit. These audiovisual clips use to last between 30 and 90s, containing a video showing short shots of the daily and quotidian life in territories conquered by the Daesh. These clips are not related with the violent films previously explained. The atmosphere show on in mujatwits is fair and optimistic. It's still a call to the Jihad but in a smoother and friendly way. These videotwits focus on civil population mostly children. In these clips, fighters are shown in a happy and comfortable way: smiling to the camera, playing with the children and participating in social activities. The clips try to give a fair image of the terrorists as protectors of the society.
4. Report: based in emphasizing the idea of caliphate of the IS as a safe, wealthy, well governed and peaceful place for its population who lives under a harmonic Sharia. Consequently, the directors will remark all the supposed benefits of the system, like the inexistence of crime, the education, the health, the legal order etc. They will also try to prove that the information published in different media that is trying to refute these claims, are obviously false. Report is created as a neutral subject, because it needs to look as true as possible in order for people to believe it is real. The latest clips of this category, are using the figure of John Cantlie, in order to get a wider diffusion. They try, in this way, to attract more families for the supposed facilities that the "Pseudo State" offers, and to avoid that part of their population to leave their land. There are also other kinds of reports that do not include any representative player. In these ones, they are using declarations of people acting like witnesses, most of the times children.
5. Documentary; these videos intend to justify the actions of Daesh, even the violations of Human Rights. The first one of these reports "Flames of War" was launched in 2014 (Al Hayat Media Center 2014). Its launching had a promotion like a Hollywood production, with trailers, banners and an advertising campaign. These documentaries are used to tell how the perfect society founded by the prophet has been corrupted by the effect of the unfaithful people and the crusaders. Thereupon they encourage people to abandon all the corrupted uses and come back to the pure origins of the Islam. Federico Aznar describes all their subjects as stories that always have a happy arcadia as a starting point, explaining the future by using the past, or, to be more accurate, to re-write the past on behalf of the future (Aznar 2014).

3.2 Cyberspace at the Service of al-Qaeda

In the last 20 years, at least 3800 audiovisual messages were issued by Al-Qaeda, a totaling 1100 h broadcasting. This has been possible thanks to the famous communication agency As Sahab, that carries all the advertising campaign of the afore mentioned group, issued in Arabic and Urdu. Al Qaeda also has a corporative magazine called Resurgence. This publication is written in English in order to reach the maximum number of readers possible. The subjects of this publication are focused in margination, prosecution and elimination of Muslims all over the world. They also make a call to their followers to destroy petrol and gas pipelines, as well as maritime routes, in order to attack Western economies (Baños 2014).

Al Qaeda and their diverse branches have given a big importance to the Internet. Within Cyberspace and binary code, terrorists have found a free and encrypted means of communication, that they can use for the broadcasting of their communications, slogans, and technical information in order to recruit more members. (Estallares y López 2011). Al Qaeda is per se, the terrorist group with the most branches. Its structure is, according to many authors, like a dispersed node that is operating with some independence. The branches have their "matrix" in Pakistan (AQ), and several franchises in various places of the world: AQMI (Al Qaeda in the Islamic Maghreb countries), AQI (Iraq), AQAP (Al Qaeda al Jihad in the Arabic peninsula), AQEA (Al Qaeda in Eastern Africa-Somalia), AQ and JI (AQ and the Jemaah Islamiyah in Indonesia). All these franchises must communicate with each other and the Internet has been the perfect tool for this communication.

The so called Technological Gap has played in favor of Al Zawahin followers, due to the enormous quantity of data flowing at high speed through the net it is impossible to analyze all of it. Security agencies were not able to intercept the September 11th attacks in the USA, as well as the subsequent ones. Al Qaeda has developed along this century, many means of communication like encrypted e-mails, steganography (sending images with hidden info) or the electronic "traffic-lights" (images previously identified by receivers that are changing the color of their background, depending on the kind of message wanted to transmit). The most popular jihadists' sites have been "Ansar Al Jihad Network" and "Al Mojahden Electronic Network". These are very striking because it is possible to interact directly with the criminal gang and get information about their latest releases, videos and claims (Bamford 2008).

3.3 Cyberspace at the Service of the Taliban

Since the Taliban started their attacks against the Afghanistan government they have started to use communication tools. In the beginning they made newspapers and periodic magazines. Today their main weapon is the digital magazine Al Somood, considered as the official publication of IS of Afghanistan and has been published since 2006. Al Emara is its official agency of production, although there are other agencies in charge of production, diffusion and edition of their videos. Between 2004 and 2012, they issued 125 messages of audio/video, achieving a total of almost 74 h of broadcasting. In April 2016, Google eliminated an app of the Taliban propaganda from its Play Store. This application was trying to increase their visibility on the net based on videos and proclamations using the Pashtun (language used in same areas of Pakistan and Afghanistan). The application tried to repeat the success of Daesh with its propaganda strategy. To this effect, the Taliban have their own channel on the service of encrypted messages Telegram as well as several web sites in different languages, besides all their regular social network profiles.

3.4 Cyberspace at the Service of al-Shabbaah

The Al Kataib foundation is producing and distributing in Somalia, high quality video-clips with the main message that Al-Shabbaab is only more element of a worldwide conflict where Islam is under threat. Their targets are the Somali people as well as the big diaspora distributed all over the world. The group is using Twitter quite often in order to deny the official declarations of the African Union. In December of 2011, they started to issue communications through an account called HSMPress that has more than 8000 followers along with their own radio station called Al-Andalus that reaches the farthest corners of Somalia. Another fact that attracts attention is their intriguing use of music. Abu Mansoor Al-Amriki, well-known leader of Al Shabbaab, got lots of followers and was the center of media attention trough a video-clip that he uploaded on the net where he was playing a rap song exalting the Jihad (De la Corte Ibáñez 2015).

3.5 Cyberspace at the Service of Boko Haram

Boko Haram, refered to by themselves as the Group of the People of Sunnah for Preaching and Jihad, which roughly means "Western Education is a sin", is the name of a terrorist group known worldwide, that is acting in Africa, especially inside Nigerian territory (Echevarría 2014). Boko Haram has been transformed in the true intercommunity Nigerian conflict, taking advantage of the rivalry between Christians and Muslims for re-fuelling their jihadist combat. In 2010, 500 people were killed by Fulani farmers in intercommunity fights in Jeji, Ratsat, Nahaua and Fulani. These crimes were promoted by Boko Haram, pushing farmers to kill with knives and other weapons. In this case, incitement messages were transmitted through mobile phones. Boko Haram is also using massive messages by video, with the support from AQMI, through their sophisticated Al-Andalus institute of communications (Zenn 2014, p. 12).

3.6 Cyberspace at the Service of Jabhat al Nusrah

The Al Nusrah branch is considered as the official arm of Al Qaeda in Syria and they also have their own personal system of communications. They have a magazine that summarizes the activities carried out in the war, entitled Monthly

Harvest, dedicating a significant space to promote the Dawwa of the organization, as well as their charity works. For example, they issued photos of members of the organization participating in a vaccine campaign against the poliomyelitis. Obviously they are trying to get the support of the local population as well as to prove that the plans of the group bring certain benefits to Syria. They carry out this kind of actions due to the fact that a lot of the local population is resisting the implantation of the ideology of Al Qaeda. They are also producing videos through an agency of communication called Himam News Agency. These media products range from reports to high quality releases. They also have another agency—Al Manarah Al Bayda Foundation for Media Production—that is in charge of distributing their most recent videos. Twitter is another of the strengths of Al Nusra communications. The fact that this group is not very well known compared with the supremacy of ISIS, entail that their messages are not observed as much.

The group has ten official accounts in that social network. Among those ten accounts, they have on the main one; the rest are merely secondary accounts. The design of the account is always quite neutral. They don't use symbols of the group or other elements that can attract too much attention. Another strategy consists in using old unusable accounts and changing their contents accurately, to avoid new policies of affiliations. Although no one can beat Daesh in audiovisual contents, Al Nusrah has shown a more efficient use.

4 Conclusion

The present chapter provided an overview, developed along functional lines, of the means by which the Internet is often utilized to promote and support acts of terrorism, in particular with respect to propaganda (including for the purposes of recruitment, radicalization and incitement to terrorism), training and financing, planning and executing such acts. Emphasis is also placed on the opportunities offered by the Internet to prevent, detect and deter acts of terrorism. Authorities will require the cooperation of telecommunications operators when undertaking electronic monitoring, wiretaps and similar electronic investigative technique. Internet-related data (e.g. customer usage) will be important evidence in many terrorism cases. In such cases, authorities should ensure that relevant data is preserved for later evidential use in proceedings.

References

Aznar F (2014) El papel de la narrativa en el terrorismo. In Aznar F, Baca E, Lázaro J (eds) La guerra contra la violencia. Triacastela, Madrid

Bamford J (2008) The shadow factory (The Ultra Secret NSA from 9/11 to the Eavesdropping on America). The Echelon Program. Doubleday

Baños P (2014) Medios y modos de comunicación de los grupos extremistas. Retrieved from https://www.esglobal.org/los-medios-que-mas-le-gustan-a-los-yihadistas/

Barrancos Larráyoz D (2014) Los community managers del terror: la propaganda online de ISIS y su ofensiva sobre Irak. Instituto Español de Estudios Estratégicos. Retrieved from: www.ieee.es/.../2014/DIEEEO82bis-2014_ISS_DavidBarrancos.pdf

Castells M (1999) La Era de la Información: Economía, Sociedad y Cultura La Sociedad Red. Siglo XXI, México

De La Corte Ibáñez L (2015) Al shabaab en el cuerno de África. Instituto Español de Estudios Estratégicos. Retrieved from: http://www.uma.es/foroparalapazenelmediterraneo/?page_id=18

De La Fuente P (2016) La propaganda de reclutamiento del Daesh a través de sus videos. Documento de opinión. Instituto Español de Estudios Estratégicos. Retrieved from www.ieee.es/Galerias/fichero/BoletinesIEEE3/2016/boletinieee1.pdf

Echevarría J (2014) El desafío terrorista de Boko Harama en Nigeria. Colección: "Grupos militantes de ideología radical y carácter violento". Instituto Español de Estudios Estratégicos. Retrieved from http://docplayer.es/18482742-El-desafio-terrorista-de-boko-haram-en-nigeria.html

Estallares y López JC (2011) Los medios de comunicación de Al Qaeda y su evolución estratégica. Instituto Español de Estudios Estratégicos. Retrieved from https://publicaciones.unirioja.es/catalogo/.../131_BorregoSevillano.pdf

Jordán J (2013) Manual de Estudios Estratégicos y Seguridad Internacional. Plaza y Valdes, Madrid

Kaplan RD (1994) The coming anarchy. The Atlantic Monthly. February

Kaldor M (2001) Las nuevas guerras. Violencia organizada en la era global. Tusquets, Barcelona

Marshall A (2015) How Isis got its anthem. The Guardian, 09-11-2014, vol. 2015, NYE

Nye Jr JS (2011) Nuclear lessons for cyber security? Strategic Studies Quarterly, Invierno, pp 18–38

Orellana J (2013) Fundamentos de narrativa audiovisual. CEU Ediciones, Madrid

Patterson LV (2002) Information operations and asymmetric warfare...are we ready?. US Army War College, Pennsylvania

Porth JS (2007) U.S. Ship To Host Multinational Experts off African Coast: Maritime safety and security are focus of African training mission. U.S. Department of State. America.gov November 2007. Retrieved from http://www.america.gov/st/washfileenglish/2007/November/20071109095818sjhtrop0.1391107.html

Sánchez Medero G (2010) La nueva estrategia comunicativa de los grupos terroristas. Revista EnfoquesVIII(12)

Seck HH (2016) ISIS may try to launch cyberattacks against US, NSA Chief Warns. Defensetech magazine. Retrieved from: http://www.defensetech.org/2016/04/05/isis-may-try-to-launch-cyberattacks-nsa-chief-warns/

Symantec (2013). Norton Security Report. Retrieved from: http://www.symantec.com/about/news/resources/press_kits/detail.jsp?pkid=norton-report-2013

The Global Information Technology Report 2015. World Economic Forum. 2015. Geneva www.weforum.org/gitr. ISBN: 978-92-95044-48-7

Torres Soriano MR (2015) ¿Es el yihadismo una ciber-amenaza? Revista de Occidente, No. 406, marzo 2015, pp 20–34

Weiman G (2004) www.Terror.Net: How modern terrorism uses the Internet. United States Institute of Peace. Retrieved from https://www.usip.org/sites/default/files/sr116.pdf

Zenn J (2014) Boko Haram's mass-kidnapping in Chibok: Shekau's gains and objectives. Retrieved from http://www.ecoi.net/local_link/276855/406124_de.html

Author Biographies

Claudio Augusto Payá-Santos is Director of the Security and Defense Studies at the Nebrija University. He is a researcher in the areas of Law, Criminology and Security; member of the Nebrija-Santander Chair on Risks and Conflict Management. Currently he is a Ph.D. Candidate.

Juan José Delgado-Morán is a Lecturer in Nebrija University in the area of Security and Defense and researcher in the areas of Law, Criminology and Security, as well as member of the Nebrija-Santander Chair on Risks and Conflict Management. Currently he is a Ph.D. Candidate.

Mythology of Cyber-Crime—Insecurity & Governance in Cyberspace: Some Critical Perspectives

Steve Wright

Abstract There was a time within even this author's memory, when there was no cyberspace, no cybercrime of note, no viruses and no anti-virus software, no hacking and no hackers. Cyber-delinquency was unknown, criminals had to do their criminality in the physical world and academic research was done in libraries not 'on-line'. The speed of banking in that far off time was pedestrian. During the Fifties, letters took weeks to arrive overseas, with anything more urgent being sent by costly telegram, over phone wires. In the intelligence world, the success of decoding Enigma and the entire field of de-encryption remained a secret, Alan Turing continued to be an unsung hero, and machine intelligence had very little acknowledged role, it was mainly human centred. In the Sixties, protest was on the streets and no-one, apart from traffic engineers, knew what networking meant. In just one lifetime, all that has changed and the pace of that change has rapidly accelerated too. The evolution of cyberspace has brought many advantages to societies once separated by distances but now able to communicate, bank, educate and socialize online and in real time. It has also brought many unanticipated dangers. Some, including radicalization, grooming, phishing, banking fraud, stalking, identity theft and denial of service attacks, are the stuff of daily news. Others, including the security and defence revolutions in military affairs, are much less discussed, despite the fact that the cyber-world originated and is firmly rooted in a military architecture of space based satellites and associated communications infrastructure. This chapter critically assesses some of the mythology of just who are the cyber bad guys, the extent to which these constructions are open to wider processes of perceptions management and the need to identify the rather more hidden agendas facilitated by emerging new capability sets in cyberspace and the so called 'internet of things.' That world is still tremendously Anglo-centric, notions of

Submitted: 8.12.16; Accepted: 26.01.17.

S. Wright (✉)
University of Leeds, Leeds, UK
e-mail: S.T.Wright@leedsbeckett.ac.uk

J.M. Ramírez and L.A. García-Segura (eds.), *Cyberspace*, Advanced Sciences and Technologies for Security Applications, DOI 10.1007/978-3-319-54975-0_13

just whose security is being protected remain contested, and we are only at the beginning of a more global debate on big data and the challenge of meaningful governance.

Keywords Crackers · Cybercriminals · Cyber-attacks · Governance · Hackers · Hactivists · ICRAC · Mythology · NSA · OSCE · Perception Management · Prism · State-Related Actors · Snowden · STOA · Targeting

1 Introduction

There was a time within even this author's memory, when there was no cyberspace, no cybercrime of note, no viruses and no anti-virus software, no hacking and no hackers. Cyber-delinquency was unknown, criminals had to do their criminality in the physical world and academic research was done in libraries not 'on-line'. The speed of banking in that far off time was pedestrian. During the Fifties, letters took weeks to arrive overseas, with anything more urgent being sent by costly telegram, over phone wires. In the intelligence world, the success of decoding Enigma and the entire field of de-encryption remained a secret, Alan Turing continued to be an unsung hero, and machine intelligence had very little acknowledged role, it was mainly human centred.

In the Sixties, protest was on the streets and few apart from traffic engineers, knew what networking meant.[1] In just one lifetime, all that has changed and the pace of that change has rapidly accelerated too. The evolution of cyberspace has brought many advantages to societies once separated by distances but now able to communicate, bank, educate and socialize online and in real time. It has also brought many unanticipated dangers. Some, including radicalization, grooming, phishing, banking fraud, stalking, identity theft and denial of service attacks, are the stuff of daily news. Others, including the security and defence revolutions in military affairs, are much less discussed, despite the fact that the cyber-world originated and is firmly rooted in a military architecture of space based satellites and associated communications infrastructure.

This chapter critically assesses some of the mythology of just who are the cyber bad guys, the extent to which these constructions are open to wider processes of perceptions management and the need to identify the rather more hidden agendas facilitated by emerging new capability sets in cyberspace and the so called 'internet of things.' That world is still tremendously Anglo-centric, notions of just whose security is being protected remain contested, and we are only at the beginning of a more global debate on big data and the challenge of meaningful governance.

[1]Yet even then there were inklings of vulnerability when according to McMillan (2012) cited in Colajanni (2016), users of the IBM mainframe CTSS system at MIT were able to access each other's passwords. Colajanni records the first software able to replicate itself amongst networked machines as Creeper which emerged in 1971; and the first destructive computer 'worm' as Morris which was released with destructive effect in November 1988.

Now the entire commercial sector is dependent on networks of machines that give instant access globally. Decisions which were once the stuff which was formally written up on parchment, are now stored 'in the cloud'; city utilities from traffic lights to energy supplies are networked via the world wide web; whilst robbery and bank heists can be accomplished remotely without coming into physical contact with the materials stolen and often at a very safe distance away, even in another country. We now have a new cultural language, a new 'surfing' culture and many daily activities are accomplished remotely in cyberspace. To anyone born in the Fifties or before, much of this new reality is akin to magic, action at a distance where goods and services can be ordered without going into shops; where children can be bullied and groomed by strangers at large in the so called 'dark web'; and transformative 'radicalisation,' can happen, online, home alone in a bedroom.

The popular stereotype of the awkward geek who can break into the computers of even the most top secret accounts, is the stuff of movies. Much less has been said about how these new capacities have been harnessed by military, security and policing agencies. Post 9/11 vast sums of money have been devoted to the business of counter-terrorism, much of which has gone into the world of cyber counter-terrorism. What is much less discussed is the extent to which these new capacities have fostered new approaches to espionage, warfare and security that go beyond the limits of the law. A crucial question behind the creation of this ultra-networked world, is one of governance. Who sets and polices the standards and what elements are creeping beyond notional human control? In security terms what are the social and political implications of creating a facility where most people can be tracked and or targeted via geo-location? Does privacy no longer have any meaning and to what extent are people actually bothered? This chapter explores some of the mythologies around this new reality and the implications of almost unbelievable processing capacities which link nearly all of those in possession of ICT devices, into a web of global surveillance.

2 Scoping the Problem: Who Are the Principle Attackers in Cyberspace?

Colajanni and Marchetti (2016) of the University of Modena and Reggio Emilia, recently summarised the main cyber attacker profiles into five categories, namely: (i) Hackers; (ii) Hacktivists; (iii) Crackers; (iv) Cybercriminals and (v) State-related actors.

Colajanni and Marchetti examined these five main attack profiles, motivations and targets for each category and found that:-

Hackers are mainly motivated by curiosity. This group had a passion for technical challenges and money was not the motivation. They are mainly interested in 'challenging systems for access.'

Hacktivists and financially supported hacktivists were mainly motivated by ideology, politics and religion. Their targets were essentially 'systems that can be misused for propaganda' and/or 'systems of opposing factions.'

Crackers are mainly motivated by a desire to vandalise or create denial of service (Dos). Their targets are mainly 'systems vulnerable to simple attacks or not protected against DoS.'

Cybercriminals are largely motivated by potential 'economic gain through data stealing, ransom, rental and sale of systems, malware, information.' Their targets are essentially 'systems that contain valuable data; systems used to make money (online stores, payment systems)' etc.

State-related actors are paid agencies and personnel authorised to seek intelligence, social control or to engage in cyber-warfare and espionage. Their targets include 'internet service providers, critical infrastructures,' the communications of foreign officials and companies, for both economic and political gain.

3 The Changing Virtual Landscapes—Policing the Interface Between Cyber Worlds and External Realities

A key challenge is deciding which category is which. Is an attack local, corporate, state orientated or terrorist inclined and how should it be quantified? Different attackers are demanding, deploying and seeking different resources and nearly all actors are masking both their identity and intent. There is a Darwinian dimension of evolving ever more difficult to detect algorithms, which are tested against major anti-virus software before being released. The work of Colajanni and Marchetti (2016) is informative here because they place such developments within an epoch from 1996–2020 characterized by two significant features, namely the rapid evolution of Internet-based services and a significant improvement in the tools required to make any cyber-attack successful.

These trends have been exacerbated by the rise of social networks; the creation of the internet of things; and the rapid evolution of portable and wearable ICT devices. Such developments have meant that millions of people have become willing to share much of their personal and professional information on line which has led to new fine grain surveillance capacities for monitoring individuals and organisations. In the workplace, employees can become the weakest link to penetrating previously confidential material, newly vulnerable to socially engineered attacks.

The internet of things has led to a massive increase of devices connected to the internet which have unanticipated vulnerabilities to cyber-attack. The staff who meticulously update and deploy anti-virus software on their laptop, are usually oblivious to the risks of the attached or networked devices such as printers and copiers.[2]

[2]Colajanni and Marchetti cite Ackerman (2016), re the Denial of Service Attack of November 2016 in the US, which was caused by a botnet infecting hundreds of thousands of devices.

Portable and wearable devices are being increasingly used for private and professional purposes. Geo-location architecture enables selective and continuous tracking. For Colajanni and Marchetti, this leads to a dissolving of the concept of the perimeter of enterprise and organizational networks—which make it far more difficult to design effective defences.

During this epoch there has been 'an industrialisation of cyber-attacks,' as the cyber-criminal ecosystem has become a 'mature economy.' They cite the industrialization processes which have driven the tools for automatic malware generation 'causing an explosion in the number of unique malwares. They claim that the current rate is half a million unique malwares every single day has been sufficient to overwhelm traditional anti-virus software.

In terms of policing such advances, the balance of this cat and mouse game is shifting in favour of the mice. The sentinel cat has to be effective 100% of the time whereas the invasive mice need only get through once to make a protection system fail. This has led to many shifts in emphasis both practical and PR. Antivirus makers have had to adopt machine learning and behavioural detection approaches and in the case of banks, simply covering up the scale of vulnerability to maintain consumer confidence. As we shall see below, this tactic has begun to crumble in the face of so called Advanced Persistent Threats (APT) which include attacks that go undetected for months or even years.

During this period of rapid change, we have seen the emergence of far more state related actors, who are characterised by both their access to astronomical financial resources and privileged access to internet service providers and critical infrastructures. Part of the modern mythology is that this paradigm shift amongst state actors was generated by the terrorist attacks of 9/11. The reality, as we shall see below, is that such capability sets were evolved in parallel with modern communication systems, often with agendas that went far beyond the limits of the law.

4 Who Are the Major Players—What Domains Do They Control?[3]

Even a cursory snapshot of news of attacks in the immediate period before this chapter was completed, illustrates the variety and complexity of the challenges. Various themes clearly emerge, apparently further evidencing the taxonomy of Colajanni and Marchetti (2016) (above). Even such a Vox-Pop sample, appears to give a flavour of one of the fastest growing security threats of our time. We find new forms of digital crime encompassing ever more areas of economic, social and political life. The news also gives us an inkling of the sheer scales of some cyber-attacks, their impacts, costs and consequences. The news informed us that in regard to:-

[3]With thanks to Colajanni (2017), Rauf (2017) for these case examples.

4.1 Cyber-Criminals

'Cyber-attacks on Talk Talk deprives 100,000 of web access' (*Guardian* 2 Dec. 2016); Yahoo says 1 Billion User Accounts were hacked', (*New York Times*, 14 Dec. 2016); no target was sacrosanct—'Cyber-attacks and Medical devices', (*New York Times*, 2 Nov. 2016); On-line banking was no longer safe—'Think your online transactions are safe?—don't bank on it', (*Guardian*, 22 Oct. 2016).

Worse still was that the most intimate areas of personal life, secreted in the private recesses of the cyber world, could rapidly become public domain, with all the attendant risks of blackmail, loss of relationships and social capital—'Adult friend Finder hacked in massive personal data breach,' (*New York Times*, 14 Nov. 2016). Cyberspace is increasingly emerging as a place for bullying without respite. 40% of American adults complaining of online abuse and 73% having witnessed it —(*Guardian* says Police forces acknowledge that existing laws not working, 12 April 2016); 'On-line abuse aimed at writers is more likely if you are female study suggests,' (*Guardian*, 13 April 2016); Is it too late to stop trolls trampling over our whole political culture?' (*Guardian*, 13 April 2016). Such realities place increasing pressure on local police forces which are ill equipped to offer adequate responses other than 'unfriending' the guilty partners. Increasingly they are being forced to become information workers, with all the specialist expertise, that role must entail.

4.2 Hactivists

A problem with attribution is that unless a state or non-state actor claims responsibility, the media will simply report news of damage by hackers; —'Yahoo says accounts 1 billion user accounts hacked in 2016, (*New York Times*, 14 Dec. 2016); 'Massive Cyberattacks blocked your favourite websites, Twitter Paypal, Amazon and others,' (*New York Times*, 21 Oct. 2016); 'Hackers Used New Weapons To Disrupt Major Websites Across US,' (*New York Times*, 21 Oct. 2016). Some of these stories centred on Domain Name Service provider Dyn, allegedly by Hacktivist groups Anonymous and New World Hackers, involving multiple denial of service attacks (DoS), again though a less protected set of security interfaces via the Internet of Things (IoT). https://en.wikipedia.org/wiki/2016_Dyn_cyberattack.

But the sources of many other attacks will always remain contested: 'After the Cyber-attack to Deutsch telecom, Germany fears Election Disruption,' (*New York Times*, 8 Dec. 2016); or 'US & Ecuador reject conspiracy claims after Assange's internet access is blocked,' (*Guardian*, 20 Oct. 2016).

4.3 State Related Actors

The news coming in regarding state actors had much larger consequences relating to the economies and even the political futures of whole countries (e.g.,) ‘Cyber-attacks Strike Saudi Arabia, Harming Aviation Agency,’ *New York Times*, 1 Dec. 2016). But the most significant story so far and a first, is the allegation, vehemently denied, that Russia leaked sensitive emails from presidential contender, Hilary Clinton at a critical moment during the US Presidential election:’Cyber-attacks were revenge for claim Russian election was rigged says Clinton,’ (*Guardian*, 17 Dec. 2016); these allegations became part of the critical aftermath of a highly polarised election result. Incoming President Donald Trump’s rubbishing of these claims was contradicted by his own intelligence officials who claimed: ‘Putin Ordered “Influence Campaign” Aimed at US Election Intelligence Report Says’, (*New York Times*, 6 Jan. 2017) and they named names—‘Hacker accused of interfering in US election’, (Guardian, 7 Jan. 2017).

But is that the full story? Ever since the attacks of 9/11, surveillance experts such as David Lyin, (Lyon 2003) have warned of an over-militarization of civil space where there is little public debate. Few have stood up to the juggernaut of security agency demands for ever widening access to public use of cyber space, Apple being one honourable exception: ‘Apple battles FBI for Privacy’ (*Guardian*, 20 Feb. 2016). Other reports opened up this dimension further—‘Privacy Campaigners fear unfettered surveillance’(*Guardian*, 12 Nov. 2016), where an NSA Whistle blower Thomas Drake is cited—‘The electronic infrastructure is fully in place—and ex post facto legalised by Congress and executive orders—and ripe for abuse under an autocratic, power obsessed president. History is just not kind here.’ Other campaigners warned of the consequences of building a firewall to protect British companies from Malicious attacks—‘Great British firewall stokes privacy fears’, (*Guardian*, 15 Sept. 2016). Their worry is that it will be used to deny freedom of speech where any unapproved sites could become designated malware.

When we discuss future governance below, it is crucial to have some deeper understanding of objective risk. Despite massively increased security expenditure with a post 9/11 doubling of MI5 budgets in the UK, Britain has nowhere near the level of terrorist attacks prevalent in the 1970s (Wright 2015). Indeed, whilst most states now actively prepare for a more permanent defensive and offensive cyber-presence; in statistical terms, the citizens of both the US and Europe are more at risk of dying from bee stings.[4] Clearly, we need to establish new rules for objectively allocating risk.

One explanation of this phenomenon has been provided by Professor Phil Taylor of Leeds University, now deceased, who spoke of ‘munitions of the mind’ (Taylor 2003). For him, modern warfare had moved into a new environment where shaping the ‘battlespace’ is a strategic exercise in ‘perception management’ involving

[4]http://www.telegraph.co.uk/news/uknews/terrorism-in-the-uk/9359763/Bee-stings-killed-as-many-in-UK-as-terrorists-says-watchdog.html.

'deception' 'operational security' and 'psychological operations'. This replacement of battlefields by 'battlespaces' means future wars are fought as much in the virtual world as in the physical world.[5] Modern security doctrine is rapidly revisiting the ancient aphorisms of Sun Tzu:

> To win one hundred victories in one hundred battles is not the acme of skill. To subdue the enemy without fighting is the acme of skill (Tzu (2009) Chap. 3, p. 115).

Increasingly, however, virtual battlespaces and real weapons are being unified into new geo-location based targeting systems, based on algorithms that work in both dimensions to produce innovative forms of armed vision. See (Hayes 2006), Hayes (2010) and Graham (2010). Indeed cyber-attacks have been identified as being one of the few threats that may legitimate a first order retaliatory nuclear strike from the US. But evidence is also mounting that the alleged stealth capacities of these ole school nuclear weapons submarines such as Trident, may themselves be vulnerable to cyber-attack.[6]

5 Mapping Impacts, Costs and Consequences

2016 proved a bumper year for academic, consultants and conferences on cyber security. But this general narrative of external cyber-enemies, is linked up to a dominant trope of ever-increasing vulnerability which looks all set to continue. This author got invitations to 'Cyber 2017', advertising with the strap line—'*Evolving norms, improving harmonisation and building resilience*,' to be held in Chatham House in June 2017.

The organisers explained that 'In light of warnings from the EU Commissioner for the Security Union, Julian King, that France, Germany and the Netherlands may be targets for hacking in the lead up to their elections this year, cyber-security remains high on the agenda of states, companies and individuals. https://www.chathamhouse.org/conferences/cyber-2017. It is symptomatic, during this period many universities set up their own cybersecurity centres including my own.[7] Others added to their expertise, by recruiting specialist staff.[8] States too during the last few years, are producing new guidelines for national cyber strategies and establishing institutional bureaucracies to further national policy. These include all US (Carter 2016) and (DoD 2015), UK (Baylon et al. 2015) and Israel, (Siboni and Assaf 2016). We are told that the US has become increasingly transparent about the need for its cyber doctrine to be more offensive identifying a need to 'project power by

---

[5]For a collection of Phil Taylor's publications and presentations, see http://media.leeds.ac.uk/papers/vp018c14.html.

[6]Borger (2016) 'trident is old technology- the brave new world of cyber warfare', *the Guardian*, 18 January.

[7]The Cybercrime and Security Innovation Centre http://www.leedsbeckett.ac.uk/csi/.

[8]Lecturer in cyber security—National College Ireland (*Guardian*, 27 Sept. 2016).

the application of force in and through cyberspace' (Lyngaas 2014). Yet little has been said in these public discourses about the current and historical role of the US National Security Agency (NSA) and how we can protect ourselves from intrusive wiretapping, even to the extent of using routine encryption, though Diffie and Landau (2010) did make a core contribution to that debate. Post-Snowden, this is a notable omission. (See below).

Nevertheless, all of these institutions, young and established are building research profiles to legitimate their expertise in delivering enhanced resilience. We are slowly developing a fuller understanding of some of the risks.

Jardine (2015) in an exhaustive quantification of current risks in cyberspace written for the 'Global commission on internet Governance, provides some fascinating quantitative data on the growing size of cyberspace and the number and characteristics of attacks. He concludes with the interesting and perceptive conclusion that proportionately it may not be as bad as we imagine since the absolute size of cyberspace is itself rapidly expanding. But there is a danger of taking such accounts as gospel, even though the proffered statistics seem incredibly precise. However, Jardine focusses exclusively on the cost of economic damage to companies and does not appear to utilize available data on state intelligence-led incursions. (As FBI Director James Comey has said.' There are two kinds of companies in the United States: those who've been hacked… and those who don't know it yet') (Oct. 2014).[9]

Nor does Jardine cover emergent military ideologies which focus on cyber-attacks on critical infrastructure, as targets in of themselves. Here the costs and consequences are far more difficult to calculate. The economic consequences for successfully switching off utilities in a large modern city for even 1 day or a week would be historic. The human costs would also be significant especially if life support systems in hospitals were involved, whereas a successful attack on the control systems of a nuclear installation could potentially be catastrophic.

5.1 *Switching off Cities*

Pioneering work on this aspect was undertaken by Urban Geographer, Professor Stephen Graham, at Newcastle University. Graham identified a new military urbanism where all citizens are tracked as potential targets, especially at borders. He examined the military innovations in using cyberspace and telecoms for this purpose, noting that 85% of all the emails in the world went through not four cities but just four buildings in the US (Graham 2010). Such permanent access to telecommunications traffic means targeted geo-location becomes relatively easy for those with the satellite access and appropriate algorithmic software.

---

[9]Cited by Colajanni (2017).

Few Cybersecurity experts have bothered with this area but it is far advanced. We are now witnessing the creation of simulation zones and mock up towns with sci-fi names such as Yodaville to test out new forms of information based warfare. What has become obvious is that 'rubbilizing' modern highly populated cities via ground troops is both expensive in terms of force protection, as well as the enormous costs attendant to rebuilding it, once hostilities cease. Graham predicts we will see the amalgamation of such 'demodernizing' capacities in the future, to target not only key individuals but also critical infrastructure. Added dangers emerge when such targeting becomes autonomous or semi-autonomous with campaign groups such as ICRAC seeking to ban such algorithmic systems,[10] and other groups of experts seeking new forms of watchdog to oversee developments in artificial intelligence which could undermine both state and personal security.[11]

5.2 *Targeting Critical Infrastructure*

Rather more attention has been paid to protecting critical infrastructure by government and cyber-security experts and rightly so. Whether enough has been done is questionable. Colajanni (2017) argues that no matter how well an organization defends itself, 'there will always be vulnerabilities to cyber attacks' which 'exploit human nature and technology—with 100+ new malware created every minute.' He also highlights a detection problem since 'in 60% of breached, data is stolen in hours but 54% of breaches are not discovered for months.'

Tariq Rauf, SIPRI's Director of Disarmament, Arms Control and Non-Proliferation, recently highlighted some of these core issues in relation to nuclear installations, in detail (Rauf 2017). These included coordinated attacks, attacks against cyber-physical systems and so called advanced persistent threats. Rauf identifies cyber security at civilian nuclear facilities as a growing concern, which he defines as 'the protection of confidentiality, integrity and availability attributes of electronic data, computer systems and processes including the protection of those computer systems, networks and other digital systems that are critical for the safe and secure operation of the facility and for preventing theft, sabotage and other malicious acts.' Rauf's analysis is quite damming since he identifies a long list of vulnerabilities highlighted by Chatham House reports, including 'lack of incident disclosure, paucity of regulatory standards, insufficient spending with developing countries at particular risk; insufficient cyber security training, a lack of key cyber security procedures amongst plant personnel, as well as cultural and skills gaps between operational personnel and cyber security personnel' (Ibid).

His conclusion was that the human factor 'is the weakest link in the cyber security defensive architecture.' This conclusion was based upon a range of

[10] Icrac.net.

[11] See Guardian 'Set up watchdog for artificial intelligence, say tech experts, 28 January 2017.

negative factors including staff using personal devices like USB sticks at nuclear facilities to transfer data and the fact that quite often computers in control rooms were left unattended.

The number of historic cyber security incidents quoted by Rauf are telling, including a 1992 incident at Ignalina nuclear power plant (Npp) where a technician decided to highlight cyber vulnerability by deliberately introducing a virus; the Slammer worm which infected the Davis-Besse Npp in 2003; the 2008 Hatch Npp where a contactor tried to install an update without formal permission, which led to an automatic 48 h shutdown, (Ibid) and many more. The well-known Fantasy writer, Terry Pratchett coined a name for this kind of episode from his years working as a Public relations Officer in a nuclear power plant—'the Fred Factor.' This is the tendency for the lowest paid manual worker or technician to be able to successfully side step any so called fail-safe mechanisms (Pratchett 2015) Rauf takes these threats seriously and proposes the need for detailed cyber security assessments which implement both hard and soft cyber security controls, by design.

Of course, targeting any states critical infrastructure is an act of war. However as Ambassador Carlo Trezza points out in another chapter of the present book, "In spite of the growing risk that cyber capacities could be used as military tools, no international agreement has been finalized so far to prevent a cyberwarfare involving sovereign state, as well as non-state actors (Trezza 2017). This is indeed true but it is also part of a mythology. Not only do certain key states already have such capacities, they are used every day to 'project power by the application of force in and through cyberspace.'as recently defined by the Pentagon.

6 The 'New' Surveillance Architectures and Agendas

Despite the focus of most commentators on cyber security on external state threats and especially non-state actors, the world's largest and most sophisticated network for focussing those attacks was for many decades, simply not reported. The postwar 1946 UKUSA intelligence sharing agreement between Britain and America, led to the formation of the National Security Agency in 1952. Amidst considerable secrecy, the NSA, colloquially known as 'no such agency' had at its core an objective to systematically penetrate all international telecommunications (See Murakami-Wood and Wright 2015).

Indeed, during the 1950s and 1960s, the entire post-office microwave network was structured to facilitate this core US mission, (Laurie 1980). Most local people thought the network was for colour TV until a very high profile so called ABC trial revealed the measures the authorities were prepared to take, to prevent any mention of so called Signals Intelligence.[12]

[12]For a personal account by the author see (Wright 2005).

One of the defendants, Campbell (1981, 1998, 2015a, b) used open sources to map a global surveillance network operated from many radomes housed at Menwith Hill, NSA's largest listening stations, based in north Yorkshire. Few took much notice until the current author produced a report for the Scientific Options Assessment panel of the European Parliament (STOA). (Wright 1998), using Campbell's work and that of New Zealand researcher Nicky Hager who had just authored a book on the New Zealand node, 'Secret Power' (Hager 1996). The debate on global surveillance, the first in European Parliamentary history, produced international outrage with feature articles emerging in most of the world's newspapers, many of which undertook their own research in an attempt to verify the allegations that America was spying on all EU members states fax and phone conversations. The US initially denied everything but other documentation emerged from the UK research NGO Statewatch, indicating that the US was also monitoring all computers and email connections too.

Campbell went on to produce an additional report for STOA, Interceptions Capability 2000 (Campbell 1999) which deconstructed the dictionary processes and mechanisms used by the NSA to perform the world's first attempt to achieve global surveillance capacities. The entire episode known as the Echelon Affaire, later became the first historical archive report by the European Library Information Service (Piodi and Mombelli 2014).

My report suggested further STOA reports be commissioned including Development of Surveillance Technology and the Risk of Abuse of Economic Information.[13] This report suggested that the NSA were using this surveillance network for economic espionage, taking an estimated €120 billion from the EU economy. This is the equivalent of a bank heist taking €1 million a week, every week for 2000 years, i.e., from when Christ was a baby.

6.1 The Snowden Revelations

Such claims were difficult to believe and despite its failure to adequately detect the 9/111 bomb attacks, subsequently the NSA was rewarded with significant extra funds and a massively expanded mission, which few adequately questioned. That was true until Edward Snowden leaked millions of documents which put flesh on the earlier claims (Greenwald 2014), (Harding 2014).

These so called Snowden revelations produced shock waves in the international system. With many official documents and formal powerpoints, Snowden was able to highlight the creation, role an d function of international snooping programmes and give them their official code designations e.g., PRISM. The Guardian was used to ensure international coverage and storage of the highly sensitive documentation

[13]http://www.europarl.europa.eu/RegData/etudes/etudes/join/1999/168184/DG-4-JOIN_ET(1999)168184_EN.pdf.

—which became known simply as the NSA Files.[14] The Guardian realized that the nature of this story required a different style of reporting as it showed step by step how the NSA had leaned on Internet service providers to give them unfettered access to their customers. Apple, AoL, Facebook, Skype, Google, Microsoft, Yahoo, Paltalk and youtube. Worse still, the files indicated that the NSA and its British partner GCHQ, were tapping the communications of their political friends and allies and had even set up a sting operation against the G7 meeting in London.

The stats revealed were astonishing. GCHQ could monitor 39 billion events every 24 h. Tapped fibre optic cables had the capacity to deliver 21 Petra bytes of data each day—or the equivalent of the entire content of the British Lending Library—192 times a day, every day. In fact the Guardian reported the US spied on 35 world leaders (*Guardian*, 25 Oct. 2013). Many people were left wondering who was not a target—a fear compounded by revelation surrounding project optic nerve. This programme involved gathering imagery from social websites like Yahoo—including sex texting worse still the incriminating material was being stored in Utah of all places and that face recognition programmes were being tested to hunt for persons who might be of future interest.[15] When President Obama formally left office, one of his last actions was to pardon Chelsea Manning for leaking images through cyberspace of US involvement in torture. It is of note that no such offer was made to Edward Snowden. His revelations of illegal NSA practises are obviously regarded as unforgiveable. But without him we would never have gained such insights into the sheer scale of US national security crimes.

7 Governance

During recent months, there have been focussed efforts via the UN GGE—(the United Nations Group of Governmental Experts on developments in the Field of Information and telecommunications, in the context of International Security) to address cyber security issues through an arms control lens.[16] The UN GGE according to Martino (2017) can be credited with establishing five working groups on the issue which have successfully outlined a global cyber-security agenda, as well as establishing the principle that international law applies to cyberspace.

During May 2016, national leaders from the G7 Summit were able to focus on cyber security and the resultant declaration started the process of agreeing consensual items relating to both geo-political and economic perspectives. These wide ranging principles supported “accessible, open reliable and secure cyberspace as one essential foundation for economic growth and prosperity.”

[14]https://www.theguardian.com/us-news/the-nsa-files.

[15]http://www.theguardian.com/world/2014/feb/27/gchq-nsa-webcam-images-internet-yahoo.

[16]I am grateful for the insights of Luigi Martino, for his research on this area of governance which this section strongly draws on (Martino 2017).

The G7 principles are very worthy and include respect for freedom, democracy, human rights as well as respecting and promoting privacy, data-protection and cyber security. They also commit the G7 to international norms in cyberspace to promote 'a strategic framework of international cyber stability consisting of the applicability of existing international law to state behaviour in cyber space' (Ibid).

Unfortunately, members of the G7 are those same members identified by Edward Snowden, whose telecommunications were systematically monitored by the NSA's dictionary based Echelon system in cyberspace to establish and potentially over rule, all their negotiating positions. Effectively breaching every one of the principles just ratified. Cite. We have been given no guarantees that the US NSA is not continuing to use cyberspace and targeted surveillance for economic espionage. So without adequate independent policing to ensure compliance, all we have is a rhetorical framework—which is according to Snowden, systematically breached several million times a day.

Martino also highlights the recent decision of the Organization for Security and Co-operation in Europe (OSCE) which calls for 'confidence building measure to reduce the risks of conflict stemming from the use of information and communication technologies' (Ibid). These are wonderful principles. However, without an independent mechanism for effectively scrutinizing even current systems, it will remain a simple wish list.

8 Conclusion

Where do we go to from here? Clearly we are at an important moment in history. Cyberspace provides unprecedented opportunities for education, empowerment and international solidarity. Through processes of crowd funding, communities can almost magically make possible which was initially seen as two difficult to even start. A humbling example whilst finishing this chapter was a young woman Rachel, who felt awful that as someone who had just moved to her home town of Reading, had his new bicycle stolen. She successfully crowdfunded a replacement bicycle.[17] Life affirming evidence of the kindness of strangers.

Some see cyberspace facilitating 'networks of outrage and hope', (Castells 2012), providing unparalleled means for effectively contesting power. The Arab Spring is cited as an example, though that interpretation is contested. One of the first responses of authoritarian regimes when challenged by people power is to clamp down on human rights defenders and social media.

This is not a new phenomenon, Reporters Without Borders (2003) have documented it since the turn of the century. Others have challenged the collusion of corporate security companies in supplying the tools for cyber intrusion to

[17]http://www.getreading.co.uk/news/local-news/note-reading-bike-theft-prompts-12498149.

oppressive regimes (The Citizen lab 2016). Privacy International has for example, consistently exposed the crimes of the Global Surveillance industry.[18]

The election of Donald Trump as US President has led to a plethora of executive orders. For examples, within days of election he was reported by CNN to be considering asking foreign visitors for social media and cell phone contacts.[19] His first meeting with British Prime Minister Teresa May, led to pronouncements on both sides to co-operate in counterterrorism and cyberwar initiatives. Within days Trump announced a further executive order 'Strengthening US Cyber Security and Capabilities'—bringing a new agenda of requirements and resources.[20] Given trump's cry of America First and a new clarion call for isolationism, it would be naïve to believe that the US will stop using its economic espionage facilities in cyberspace because of the G7 and OSCE resolutions.

Similarly, the leadership of a post-Brexit UK will perhaps not think twice of using its considerable interception capabilities in cyberspace, for making more informed deals outside of the EU, or even with US assistance, gathering information to help with negotiating the actual terms of Brexit in terms of trade. But emerging anti-muslim and xenophobic tendencies coupled with a yearning for the US to return to providing torture as a government service, may be a bridge too far even for us.

Now more than ever, we need to support the activities of NGO's such as Amnesty, Huridocs, Privacy International, EPIC, the electronic Frontier Foundation and all those such as Statewatch, ACLU and Big Brother Incorporated, that help us understand and challenge comping quantum shifts in state acquisition of new powers to control cyberspace.

References

Ackerman S (2016) Major cyber-attack disrupts internet services across Europe and US, The Guardian, 26 Oct

Baylon C, Brunt R, Livingstone D (2015) Cyber security at civil nuclear facilities—understanding the risks. Royal Institute of International Affairs, Chatham House, London

Borger J (2016) Trident is old technology-the brave new world of cyber warfare, the Guardian, 18January

Campbell D (1981) Phone tappers & the security state, Report No 2, London, New Statesman

Campbell D (1998) 'Tip for tap', The Guardian, 10 Sept, London

Campbell D (1999) Interception capabilities 2000, Brussels, European Parliament, 2000. http://www.cyber-rights.org/interception/stoa/interception_capabilities_2000.htm

[18]https://www.privacyinternational.org/reports/research-reports.

[19]http://edition.cnn.com/2017/01/29/politics/donald-trump-immigrant-policy-social-media-contacts/index.html.

[20]http://foreignpolicy.com/2017/01/27/draft-trump-order-kicks-the-can-down-the-road-on-cybersecurity/.

Campbell D (2015a) GCHQ & me—my life unmasking British Eavesdroppers. The Intercept, https://theintercept.com/2015/08/03/life-unmasking-british-eavesdroppers/. Accessed 27 Sept 2015

Campbell D (2015b) http://www.duncancampbell.org/PDF/nautilus%20report.pdf

Carter A (2016) 2017 defense posture statement—taking the long term view, investing in the future. Dod, USA

Castells M (2012) Networks of outrage and hope—social movements in the internet age. Polity Press, Cambridge, UK

Colajanni M, Marchetti M (2016) Cyber attacks and defenses: current capabilities and future trends, Keynote paper presented to 'Project pattern (Political affairs and technological transformations evolution and relevance For NATO strategy) Workshop,' 'Opportunities and challenges of technological change,' Foundazione Bruno Klessler (FBK), Trento, 30 Nov–1 Dec, 2016

Colajanni M (2017) The present & future of cyber security. Presentation to the ISODARCO Winter School, Andalo, Italy, 9 Jan

Diffie W, Landau S (2010) Privacy on the line: the politics of wiretapping and encryption. Cambridge, MIT Press, Mass, London

Department of Defense (2015) The DoD cyber strategy, DoD, Washington

Graham S (2010) Cities under siege—the new military urbanism. Verso, London, UK

Greenwald G (2014) No place to hide: Edward Snowden, the NSA, and the US surveillance state. Metropolitan Books/Henry Holt, New York, USA

Hager N (1996) Secret power: New Zealand's role in the international spy network. Nelson, Craig Potton, New Zealand. http://www.nickyhager.info/Secret_Power.pdf

Harding L (2014) The Snowden files,—the inside story of the world's most wanted man. Guardian-Faber Publishing, London

Hayes B (2006) Arming big brother—the EU's security research programme. Statewatch/Transnational Institute (TNI), London/Amsterdam. www.statewatch.org/analyses/bigbrother.pdf

Hayes B (2010) Neoconopticon—the EU security industrial complex. Statewatch/Transnational Institute (TNI), London/Amsterdam. http://www.statewatch.org/analyses/neoconopticon-report.pdf

Jardine E (2015) Global space is safer than you think—real trends in cybercrime, global commission on internet governance, Paper No 16, Centre for International Governance Innovation and Chatham House, UK. https://www.cigionline.org/sites/default/files/no16_web_3.pdf

Laurie P (1980) Beneath the city streets—a private enquiry into the nuclear pre-occupations of government. Panther Books, UK

Lyngaas S (2014) New cyberdoctrine shows more offense, transparency. FCW, 24 Oct

Lyon D (2003) Surveillance after September 11. Polity Press, Cambridge, UK

Martino L (2017) Cyber weapons, arms control and international legal framework: utopic ideal or realistic opportunity?—a short review of international and regional activities. Presentation to the ISODARCO Winter School, Andalo, Italy, 8–15 Jan

McMillan R (2012) The World's first computer password, it was useless too. https://arstechnica.com/tech-policy/2012/01/the-worlds-first-computer-password-it-was-useless-too/

Murakami-Wood D, Wright S (2015) Before & after Snowden. Surveill Soc 13(2):1–7

Piodi F, Mombelli I (2014) The Echelon affaire—the EP and Global interception system 1998–2002, European Parliament History Series, No 1. Publication Office of the European Union, Luxembourg

Pratchett T (2015) A slip of the keyboard. Corgi Books, UK

Rauf T (2017) Cyber security for civilian nuclear facilities. Presentation to the ISODARCO Winter School, Andalo, Italy, 8–15 Jan

Reporters Without Borders (2003) The internet under surveillance—obstacles to the free flow of information on-line. Reporters Without Borders, France

Siboni G and Assaf O (2016) Guidelines for a national cyber strategy, Memorandum No 153, tel Aviv Institute for National Security Studies, March
Taylor PM (2003) Munitions of the mind: war propaganda from the ancient world to the nuclear age. Manchester University Press, UK
Trezza C (2017) A negotiation on cyberwarfare. Presentation to the ISODARCO Winter School, Andalo, Italy, 8–15 Jan
The Citizen Lab (2016) Citizen researchers uncover new cyber espionage operation targeting Syrian opposition 2 August, Monk School of global affairs. University of Toronto, Canada
Thomas D, Loader B (2000) Cybercrime. Law enforcement, security and surveillance in the information age. Routledge, London
Tzu S (2009) The art of war (Translation by Griffiths SB). Watkins Publishing, London, UK
Wright S (1998) An appraisal of the technologies of political control: interim STOA Report (PE 166.499). European Parliament, Directorate General For Research, Directorate A, The STOA programme, Luxembourg. Reprinted at http://aei.pitt.edu/5538/
Wright S (2005) The Echelon trail: an illegal vision. Surveill Soc 3(2/3):198–215
Wright S (2015) Watching them watching us. Int J Commun Ethics 12(3–4):47–57

Author Biography

Dr Steve Wright (B.Sc, PhD, FRSA, Fellow of the Higher Education Academy) is currently a Reader in Applied Global Ethics, at Leeds Beckett University in the UK, where he teaches undergraduate and postgraduate modules on peace and conflict. He has researched new technologies of political control for over four decades, starting as a postgraduate at Lancaster University's Richardson Institute and continuing from 1984–1989, as Head of Manchester City Council's Police Monitoring Unit, where he learnt much about public opinion formation and policy change. Wright went of to chair the influential NGO, Privacy International for several years.

Part III
Cyberwarfare

The Tallinn Manual and Jus ad bellum: Some Critical Notes

Sonia Boulos

Abstract This essay critically analyzes certain aspects of "The Tallinn Manual on the International Law Applicable to Cyber Warfare". It addresses specifically the Manual's rules on the applicability of *Jus ad bellum* to cyber-attacks. The essay focuses on two particular aspects of this inquiry: the test or formula for classifying a cyber-attack as an "armed attack", and the applicability of the right to self-defence to cyber-attacks by non-State actors. While drafters of the Manual present it as a humble *lex lata* project, the essay suggests that the Manual's occasional deviation from the jurisprudence of the ICJ coupled with a possible lowering of the "armed attack" threshold, could lead to profound alterations of the landscape of future conflicts contrary to the declaration of its drafters.

Keywords Cyber-attacks · UN charter · Use of force · Armed attack · Scale and effects · Right to self-defence · Non-State actors · Rules of attribution

1 Introduction

The interest in cyber warfare has risen sharply in the last decade with the 2007 cyber-attack against Estonia. The three weeks long attack involving Russian hackers resulted in disabling the websites of the Estonian presidency and its parliament; government ministries; political parties; leading newspapers; the country's biggest banks; and firms specializing in communications of government ministries (Guardian 2007). A year later, Georgia's suffered the first ever combined kinetic and cyber-attack launched by Russia (Atlantic Council 2014). In 2010, it was

Submitted: 2.12.16; Accepted: 9.12.16.

S. Boulos (✉)
Nebrija University, Madrid, Spain
e-mail: sonia.boulos@gmail.com

J.M. Ramírez and L.A. García-Segura (eds.), *Cyberspace*, Advanced Sciences and Technologies for Security Applications, DOI 10.1007/978-3-319-54975-0_14

assumed that the US, with the help of Israel, was responsible for the Stuxnet virus launched against Iran targeting the uranium enrichment facility at Natanz, causing Iran's centrifuge machines to spin out of control (Guardian 2012). More recently, in 2014, North Korea was accused of staging a cyber-attack against Sony Pictures Entertainment. The attack paralyzed, temporarily, the computers of the company. Hackers also deleted files from hard drives; uploaded unreleased films to the Internet; and leaked sensitive personal information regarding thousands of employees (Washington Post 2014).

The Estonia cyber-attack prompted the Tallinn based NATO Cooperative Cyber Defence Centre of Excellence to invite a group of experts and charge them with preparing a study on the Applicability of international law to cyber conflicts and cyber warfare. Twenty international law scholars and experts were recruited for the task. It took the participants three years to finish their work in the form of a manual (Schmitt 2013). In April Schmitt (2013), it was published under the title "The Tallinn Manual on the International Law Applicable to Cyber Warfare".

The aim of this essay is to critically analyze the Manual's rules on the applicability of *Jus ad bellum* to cyber-attacks. It focuses on two particular aspects of this topic: the test or formula for classifying a cyber-attack as an "armed attack", and the applicability of the right to self-defence to cyber-attacks by non-State actors.

2 The Tallinn Manual: A Quick Overview

The Manual was published after three years of work that took place between 2009–2012 involving twenty renowned independent international law scholars and practitioners: the Hereinafter 'Group of Experts'. The drafting process also included consultation with information technology specialists. The Group of Experts was led by Professor Michael N. Schmitt, chairman of the international law department at the United States Naval War College, who also served as the project director.

The Manual begins with presenting itself as a *lex lata* project focusing on existing international law norms as opposed to a *lex ferenda* exercise aimed at tailoring future legislation on the subject of cyber warfare. The drafters of the Manual clearly refute the assumption that general principles of international law cannot be applied to cyber warfare, therefore, new treaty law is needed. One of the most notable features of the Tallinn Manual is its distinction between *Jus in Bello* and *Jus ad bellum* (Schmitt 2013). The manual makes a clear distinction between the international law that regulates the use of force, and international law that regulates the conduct of hostilities. In addition, the Manual addresses topics such as sovereignty, State responsibility, and the law of neutrality (Schmitt 2013). In sum, the Manual sets forth ninety five 'black-letter rules' governing cyber warfare conflicts that apparently reflect international law as it exists today. Each rule is supplemented by an extensive commentary explaining treaty law or customary law behind it. Some commentaries outline certain disagreements within the Group of

Experts on the interpretation of international law norms in relation to the rule (Schmitt 2013).

Although the Manual is a non-binding document and represents only the opinions of its drafters, it is the first serious attempt to tackle cyber warfare from an international law perspective. Prior to its publication, the Manual was peer-reviewed by thirteen international legal scholars to enhance its academic credibility. Furthermore, three organizations provided observers to the process. NATO through its Allied Command Transformation, sent observers to provide the perspective of a multinational user of the Manual. The US Cyber Command's representative offered the perspective of an operationally mature entity. The International Committee of the Red Cross was invited to observe and participate in the proceedings for being the guardian of international humanitarian law. The observers participated both in the discussions and in the drafting of the Manual. However, their consent was not necessary to achieve the unanimity required for adoption of a Rule.

3 Cyber War and the Use of Force

Can cyber-attacks or hacking activities amount to illegal use of force? If the answer is positive, can they amount to an 'armed attack'? To answer these questions the Manual consults the advisory opinion of the International Court of Justice (ICJ) on *the Legality of the Threat and Use of Nuclear Weapons*. In this specific opinion, the ICJ stated that the provisions of the United Nations Charter (the Charter, UN 1945) on the use of force do not refer to specific weapons, therefore they apply to any use of force, regardless of the weapon employed. Based on this statement, the Manual concludes that the Charter provisions on the use of force are equally applicable to cyber warfare. Thereafter, the Manual highlights the distinction between the term 'use of force' and the term 'armed attack'. To understand the importance of this distinction one needs to go back to two key provisions of the Charter that regulate the use of force in international law.

The first article, Article 2(4), states:

> all Members shall refrain in their international relations from the threat or use of force against the territorial integrity or political independence of any state, or in any other manner inconsistent with the Purposes of the United Nations (Charter, Article 2(4)).

The second article is Article 51, according to which:

> [n]othing in the present Charter shall impair the inherent right of individual or collective self-defence if an armed attack occurs against a Member of the United Nations, until the Security Council has taken measures necessary to maintain international peace and security (Charter, Article 51).

The distinction between the two terms is of a great importance because it dictates, a priori, different permissible measures available to States to respond to unlawful use of force against them. While the use of force is prohibited under the Charter, only a use of force that qualifies as an 'armed attack' triggers that right to self-defence under international law.

In attempting to sketch the differences between the term 'use of force' and the term 'armed attack' the Manual relies heavily on the jurisprudence of the ICJ. The Manual surveys the landmark case of Nicaragua v. The Unites States (the Nicaragua case), where the ICJ clarified some basic concepts regarding the legality of the use of force in international law. While the Charter lacks a definition of the term "armed attack", the ICJ stipulated that it is "necessary to distinguish the most grave forms of the use of force (those constituting an armed attack) from other less grave forms" (ICJ 1986, para. 191). The Standard of evaluation adopted by the ICJ is the 'scale and effect' standard (ICJ 1986, para. 195). An example of a prohibited 'use of force' that *does not* amount to an 'armed attack' is the indirect support of subversive or terrorist armed activities within another State. This includes arming and training of armed groups and not necessarily the mere supply of funds.

Articles 4(2) and 51 of the Charter and the ruling of the ICJ (1986) in the *Nicaragua Case* are all reflected in three Rules of the Manual. The first Rule is Rule 10, which states that:

> [a] cyber operation that constitutes a threat or use of force against the territorial integrity or political independence of any State, or that is in any other manner inconsistent with the purposes of the United Nations, is unlawful (Manual Rule 10).

Rule 11 states:

> a cyber operation constitutes a use of force when its scale and effects are comparable to non-cyber operations rising to the level of a use of force." Finally Rule 13 states "A Slate that is the target of a cyber operation that rises to the level of an armed attack may exercise its inherent right of self-defence. Whether a cyber operation constitutes an armed attack depends on its scale and effects (Manual Rule 11).

The Manual focuses first on the lower threshold, i.e., the 'use of force'. Looking at the drafting history of the Charter, the Group of Experts excluded economic warfare from the scope of the concept. In their next move, they tried to make an analogy with other non-kinetic or kinetic attacks that the international community would classify as uses of force. They identified the following eight factors that could help in assessing the nature of a cyber-attack prior to determining if it constitutes a use of force.

1. ***Severity***: operations resulting in physical harm to individuals or property will qualify as a use of force whereas those generating mere inconvenience or irritation will not.
2. ***Immediacy***: Cyber operations the produce immediate harm are more likely to be classified as a use of force compared to cyber operations that take weeks or months to achieve their intended goal.

3. ***Directness***: This factor focuses on the chain of causation between the cyber operation and the harm suffered. Cyber operations in which the cause and the harmful effects are clearly linked are more likely to be classified as uses of force.
4. ***Invasiveness***: This factor focuses on the degree to of intrusion by cyber operations into the target State or its cyber systems, in a manner that jeopardize its interests. For example, invading a military system is more invasive compared to invading a university network. The degree of invasiveness is taken into consideration in determining if a given cyber-operation meets the threshold required to be classified as a use of force.
5. ***Measurability of effects***: The consequences of cyber operations are in general less apparent than those resulting from kinetic attacks. Therefore, cyber operations that result in more quantifiable and identifiable a set of consequences are more likely to be classified as a use of force.
6. ***Military Character***: Cyber operations that have military character are more likely to be classified as a use of force.
7. ***State Involvement***: The extent and degree of involvement of states in cyber-attacks against other states is one of the factors to be considered in classifying cyber operations. A closer involvement by a State increases the likelihood to classify the cyber operation as a use of force.
8. ***Presumptive legality***: If a cyber operation is not prohibited by treaty law or international customary law, it is presumed legal. Cyber operations that are presumed legal are less likely to be classified as a use of force.

As for the threshold required for classifying an attack as an 'armed attack', the Manual stipulates that this threshold presupposes at least a use of force in the sense of Article 2(4) of the Charter. However, not every use of force meets the threshold of an armed attack; the scale and effects required exceed those required for the use of force.

The Manual clearly dedicates more efforts for articulating guidelines to identify cyber-attacks amounting to a 'use of force' rather than suggesting guidelines for identifying an "armed attack". Guidance is more needed for identifying the latter due to the consequences of such finding, i.e., allowing the Victim State to respond by using military force. Even the eight criteria elaborated in the Manual are too elusive and broad. The consideration of the eight criteria could lead to conflicting results when assessing a cyber-attack depending on the weight that is given to each or some of them.

Nguyen (2013) identifies three main approaches for assessing whether cyber-attacks cross the threshold of a 'use of force' or of an 'armed attack'. The first approach is the instrument-based approach; it focuses on the physical characteristics of the weapon used to perpetrate a given attack. If the weapon possesses the physical characteristics traditionally associated with coercive military operations, then the attack is more likely to cross the threshold of 'use of force' or even of an 'armed attack'. This approach reflects a textual reading of chapter VII of the Charter. Chapter VII authorizes the Security Council to "determine the existence of any threat to the peace, breach of the peace, or act of aggression and shall make recommendations, or decide what measures shall be taken" (Article 39). The Chapter allows for two types of measures to be employed: measures 'not involving the use of armed force' and measures that include the use of such force. The first type of measures includes, *inter alia*, "complete or partial interruption of economic relations and of rail, sea, air, postal, telegraphic, radio, and other means of communication, and the severance of diplomatic relations" (Article 41 of the Charter). The second type includes *inter alia* measures "by air, sea, or land forces as may be necessary to maintain or restore international peace and security. Such action may include demonstrations, blockade, and other operations by air, sea, or land forces of Members of the United Nations" (Article 42). If this approach is to be followed it would lead to the conclusion that a cyber-attack cannot constitute an armed attack (Nguyen 2013). This approach is highly rigid, outdated and fails to accommodate the realities of the world we live in. It is also inconsistent with the ICJ advisory opinion which refused to limit the application of the Charter's provisions on the use of force to conventional weapons.

The second approach that Nguyen (2013) identifies is 'the target-based' or 'strict liability' approach. According to this approach, any cyber-attack that targets critical national infrastructure is ought to be considered an armed attack due to the severe consequences that could result from such an attack. States may designate different institutions as critical structures. For example, in the United States, the Critical Infrastructures Protection Act of 2001 defines critical infrastructure as

> systems and assets, whether physical or virtual, so vital to the United States that the incapacity or destruction of such systems and assets would have a debilitating impact on security, national economic security, national public health or safety, or any combination of those matters (US Code 2001).

At the European Union level, in 2004 The European Commission prepared a comprehensive strategy for the protection of critical structures titled '*The European Programme for Critical Infrastructure Protection*'. The plan defined 'critical structures' as those consisting of:

> physical and information technology facilities, networks, services and assets which, if disrupted or destroyed, would have a serious impact on the health, safety, security or economic well-being of citizens or the effective functioning of governments in the Member States. Critical infrastructures extend across many sectors of the economy, including banking and finance, transport and distribution, energy, utilities, health, food supply and communications, as well as key government services. Some critical elements in these sectors are not strictly speaking 'infrastructure', but are in fact, networks or supply chains that support the delivery of an essential product or service. For example, the supply of food or water to our major urban areas is dependent on some key facilities, but also a complex network of producers, processors, manufacturers, distributors and retailers (European Commission 2004, final).

In December 2008, The European Council issued a directive on the identification and designation of European critical infrastructures and the assessment of the need to improve their protection. The directive defined 'critical infrastructure as:

> asset, system or part thereof located in Member States which is essential for the maintenance of vital societal functions, health, safety, security, economic or social well-being of people, and the disruption or destruction of which would have a significant impact in a Member State as a result of the failure to maintain those functions (European Council 2008).

The danger of the 'target based approach' lies in its potential broadness. In the absence of clear international criteria for the designation of infrastructures, States have full discretion in adopting criteria of designation, and the criteria adopted could be broad to include a wide range of institutions as reflected in the criteria advanced by the EU.

The third approach identified by Nguyen is the effects-based approach. This approach focuses on the effects produced by cyber-attacks. It looks at factors such as severity, immediacy, and directness of harm in deciding whether the consequences of a cyber-attack justify qualifying it as an "armed attack". Accordingly, a cyber-attack that produces physical destruction comparable to those caused by kinetic attacks is more likely to qualify as an armed attack.

The effects-based approach includes two types of attacks. The first type includes cyber-attacks that causes actual physical destruction, for example one that causes two trains to collide, resulting in massive destruction, death, and injuries. The second type includes cyber-attacks that target civilian or military infrastructures causing functional damage not accompanied by physical destruction. However, the impact of the functional damage is similar to the impact of physical destruction. For example, a cyber-attack targeting a power plant could shut it down without physically destroying it causing the same impact that results from physical destruction of the facility. Nguyen (2013) claims that states tend to avoid the use of force to respond to cyber-attacks not involving actual physical destruction. For example, the Stuxnet cyber-attack against Iran produced physical damage to the Iranian nuclear facility comparable to that caused by the 1981 Israeli air strikes that destroyed nuclear facilities in Baghdad. Israel's attack on the Iraqi nuclear facility was condemned by the Security Council for violating international law (Crawford 2012). However, Nguyen highlights that few States have explicitly condemned the Stuxnet cyber-attack as an armed attack or an unlawful use of force.

The Manual clearly adopts the effects-based approach. According to the Manual, any use of force that "injures or kills persons or damages or destroys property would satisfy the scale and effects requirement (p. 56). At the other end of the spectrum, the Group of Experts believed that cyber operations for intelligence gathering, or cyber theft, or cyber operations involving a brief interruption of non-essential cyber services do not reach the threshold of an 'armed attack'. However, a cyber operation directed against major systems of a State's critical infrastructure causing severe effects would qualify as an armed attack (Schmitt 2013).

The assertion that any death or physical destruction amounts to an "armed attack" is problematic, because it could be understood that even one death or small scale destruction can cross the threshold of an "armed attack". Constantinou summarizes the requirements for an armed attack as follows:

> [A]rmed attack implies an act or the beginning of a series of acts of armed force of considerable magnitude and intensity (i.e., scale) which have as their consequence (i.e., effects) the infliction of substantial destruction upon important elements of the target State namely, upon its people, economic and security infrastructure, destruction of aspects of its governmental authority, i.e., its political independence, as well as damage to or deprivation of its physical element namely, its territory (Constantinou 2000, pp. 63–64)

Her use of the terminology "considerable magnitude" or "substantial destruction" suggests that isolated cases of death or injury or isolated cases of physical destruction might not cross the "armed attack" threshold. Even scholars who believe that the threshold gap between the "use of force" and an "armed attack" should not be exaggerated, they are still hesitant to qualify isolated cases of injury or death as an armed attack. For example, Dinstein that an armed attack "must leave behind a trail of human casualties or ample destruction of property" (2013, p. 279). Therefore, it seems that the Manuals could lower substantially the threshold of an armed attack if physical injury or destruction are categorically classified as an "armed attack" regardless of their actual scale.

4 Cyber-Attacks and Non-State Actors

Rule 6 of the Manual states "A State bears international legal responsibility for a cyber operation attributable to it and which constitutes a breach of an international obligation" (2013, p. 29). For example, a cyber-attack carried out by State's intelligence agencies or by a private contractor could qualify as a use of force if the attack is attributable to the State based on the law of State responsibility. Apparently, the Manual relies on the Draft Articles prepared by the UN Law Commission on Responsibility of States for Internationally Wrongful Acts (UN 2001). According to the Draft Articles, a State can be held responsible for an internationally wrongful act when the act is attributable to the State under rules of international law. In Article 8, the Draft Articles suggest that actions of non-state actors could be attributed to a State only when the former act "on the instructions of, or under the direction or control of, that State". Article 11 suggests that "Conduct which is not attributable to a State… shall nevertheless be considered an act of that State under international law if and to the extent that the State acknowledges and adopts the conduct in question as its own".

However, the Manual creates a clear-cut distinction between the law of responsibility and the right to exercise self-defence on the territory of another State. In other words, according to the Manual the right to self-defence could be exercised in certain cases on the territory of another State even when the latter is not legally responsible for the armed attack against the responding State. For example, the majority of the Group of Experts held the view that self-defence is permissible on the territory of another State not only when the armed attack is initiated or controlled by the State, or subsequently adopted by it, but *also* when the State is *unable or unwilling* to take effective measures to repress cyber-attacks originating from its territory by non-State actors. The "unable or unwilling" significantly expands the rules of attribution adopted by the Draft Articles. Beyond the tests of "guidance", "control" or "subsequent adoption", Article 10 of the Draft Articles states that conduct of an insurrectional movement can be considered as an act of the State only if the movement becomes the new Government of the State, or if it succeeds in establishing a new State in part of a pre-existing State. There is nothing in the Draft Articles to suggest that the "unwilling or unable" test suffices for attributing the actions of non-State actors to States and holding the latter responsible for those acts. The minority view within the Group of Experts attested the use force in self-defence on the territory of a State to which the armed attack is not attributable under traditional rules of attribution. The minority view is consistent with the jurisprudence of the ICJ. The test of "unable or unwilling" was rejected by the majority opinion in the Case concerning Armed Activities on the territory of the Congo, according to which the unwillingness or inability of a State to deal with irregular forces on its territory is insufficient to create a right in self-defence against that State. The official position of the ICJ remains that the right to self-defence can be exercised against a non-State group on the territory of another State only when the group's actions can be attributed to a State under the laws of attribution. Instances where the ICJ could attribute the actions of non-State actors to a State include the sending by or on behalf of a State of armed groups, which carry out acts of armed force against another State of such gravity as to amount to an armed attack.

More so, the Manual addresses independently the broader question whether the right to self-defence is applicable to attacks by non-State actors. While the Manual recognizes that traditionally Article 51 of the UN Charter and customary international law on the right to self-defence applied exclusively to inter-State conflicts, it claims that the 9/11 attacks have changed the law. In 2001 the international community viewed the said attacks against the United States as an "armed attack" triggering the right of self-defence. This is implied in Security Council Resolution 1368, which recognized "the inherent right of individual or collective self-defence in accordance with the Charter" (UN Security Council 2001) in relation to 9/11 attacks. Dinstein (2005) argues that there is nothing in the language of Article 51 of the Charter that limits the right to self-defence to inter-States conflicts. He further emphasizes that not only the Security Council has recognized the right to self-defence against non-State actors, but the 9/11 attacks triggered other international organization to recognize such a right. This includes the first invocation ever of Article 5 of the North Atlantic Treaty by the Atlantic Council on collective self-defence (NATO 1949),[1] and the adoption of a resolution by the Ministers of Foreign Affairs, acting as an Organ of Consultation, in application of the 1947 Inter-American Treaty of Reciprocal Assistance, stating that "these terrorist attacks against the United States of America are attacks against all American States" (OAS 2001, para. 1).

[1]According to Article 5 an armed attack against one or more of the Allies in Europe or North America "shall be considered an attack against them all".

Other scholars present an entirely different reading of the above-mentioned Resolutions. For starters, Resolution 1368 requires all States to hold accountable all those responsible for "aiding, supporting or harboring" terrorists (UN Security Council 2001, p. 3). One could reasonably argue that the international support for operation "Enduring Freedom" in Afghanistan does not constitute new grounds for invoking the right to self-defence, rather it is a case where the actions of non-State actors i.e., Al Qaida were attributed to the de facto Government of Afghanistan i.e., the Taliban Regime. This reading could be inferred, for example, from the view expressed by NATO Secretary General at the time, Lord Robertson, who declared on behalf of NATO that "the individuals who carried out these attacks were part of the world-wide terrorist network of Al-Qaida, headed by Osama bin Laden and his key lieutenants and *protected by the Taliban*" (NATO 2001, emphasis added). O'Connell claims that the mere harboring of terrorists is not sufficient for attributing their acts to the harboring State. However, in the case of the Taliban Government, the latter "developed such close links to the known terrorist organization al Qaeda that it became responsible for the acts of al Qaeda" (O'Connell 2002, p. 901). Murphy (2002, p. 50) argues that Afghanistan was legally responsible for the actions of Al-Qaeda because it allowed the latter to operate from its territory despite knowledge of its intent to project force abroad, and because it failed to extradite Al-Qaeda operatives, thereby adopting the latter's conduct as its own.

It should be emphasized that after 9/11, the ICJ had more than one occasion to revisit its previous ruling on the inapplicability of self-defence to non-State actors, nevertheless, it decided not to depart from its traditional position. For example, in the Israeli Wall Advisory Opinion of 2004, the ICJ reiterated its view that the right to self-defence applies between States only (ICJ 2004). In 2005 the ICJ had a second opportunity to revisit its position in case of the Armed Activities on the Territory of the Congo. In this case the ICJ refused to acknowledge that Uganda had a legitimate claim to self-defence in relation to attacks by irregular forces (the Allied Democratic Forces) operating from the DRC, because it found no satisfactory proof of direct or indirect involvement of the DRC in these attacks. It should be further emphasized that even those who call for applying the right to self-defence to non-State actors, limit their argument to large scale attacks. For example, in his dissenting position in the Armed Activities case, judge Simma argued that:

> Security Council resolutions 1368 (2001) and 1373 (2001) cannot but be read as affirmations of the view that l*arge-scale attacks* by non-State actors can qualify as "armed attacks" within the meaning of Article 51 (ICJ 2005, para. 11, emphasis added).

A study conducted by The Royal Institute of International Affair suggests that when an attack is carried out by non-State actors, it must be a large-scale attack for the victim state to be able to invoke the right to self-defence (Wilmshurst 2005). This serious limitation is not adopted or accounted for by the Manual.

Furthermore, it seems odd that the Manual does not adopt the view of the ICJ on the non-applicability of the right to self-defence to non-States actors, but at the same time the Manual states that Article 2(4) of the Charter does not apply to non-State actors unless they are attributable to a State pursuant to the law of State responsibility. The Manual goes on to stipulate that in such a case, only the State would be found in violation of international law on the prohibition of the use of force.

5 Conclusion

There is no doubt that the Manual raises both interesting and important points on the applicability of *Jus ad bellum* to cyber-attacks. It is also clear that the drafters of the Manual were very cautious not to depart from a *lex lata* approach to their project. However, certain deviations from the jurisprudence of the ICJ, especially on the applicability of the right to self-defence to non-State actors, coupled with a possible lowering of the "armed attack" threshold, are not as modest as they may seem. In fact, they could lead to profound alterations of the landscape of future conflicts.

Cyber-attacks by non-State actors are easier to camouflage. They require significantly lower level of structural organization, of material and human resources, of a time framework, and of a geographical space to be occupied for their execution. This, in turn, means that States have significantly more difficulties in spotting and intercepting cyber-attacks compared to their ability to identify and prevent kinetic attacks by non-State actors. If the rules of that Manual were to apply, they would significantly increase the potential of using force against a large number of States. The Manual might not create new rules, but stretching existing ones can still have to a dramatic impact on the map of future conflicts.

References

Atlantic Council (2014) Russian cyber strategy and the war against Georgia, 17 Jan 2014. Retrieved from: http://www.atlanticcouncil.org/blogs/natosource/russian-cyber-policy-and-the-war-against-georgia. Accessed 24 Oct 2016

Constantinou A (2000) The right of self-defense under customary international law and article 51 of the UN charter. Sakkoulas, Athens

Crawford J (2012) Brownlie's principles of public international law, 8th edn. Oxford University Press, Oxford

Dinstein Yoram (2005) War, aggression and self-defence. Cambridge University Press, Cambridge

Dinstein Yoram (2013) Cyber war and international law: concluding remarks at the 2012 Naval war college international law conference. Int Law Stud 89:276–287

European Commission (2004) The European programme for critical infrastructure protection. COM/2004/0702

European Council (2008) Council directive 2008/114/EC. Retrieved from: http://eur-lex.europa.eu/LexUriServ/LexUriServ.do?uri=OJ:L:2008:345:0075:0082:EN:PDF. Accessed 11 Nov 2016

Guardian (2007) Russia accused of unleashing cyberwar to disable Estonia, 17 May 2007. Retrieved from: https://www.theguardian.com/world/2007/may/17/topstories3.russia. Accessed 4 Nov 2016

Guardian (2012) US was 'key player in cyber-attacks on Iran's nuclear programme', 1 June 2012. Retrieved from: https://www.theguardian.com/world/2012/jun/01/obama-sped-up-cyberattack-iran. Accessed 8 Dec 2016
ICJ (1986) Military and paramilitary activities in und against Nicaragua (Nicaragua v. United States of America). Merits, Judgment, ICJ Reports, p 14
ICJ (2004) Legal consequences of the construction of a wall in the occupied Palestinian Territory. Advisory Opinion, ICJ Reports, p 13
ICJ (2005) Armed activities on the territory of the Congo (Democratic Republic of the Congo v. Uganda). Judgment, ICJ Reports, p 168
Murphy SD (2002) Terrorism and the concept of "Armed Attack" in Article 51 of the UN Charter. Harvard International Law Review 43:41
NATO (1949) Collective Defence—Article 5. Retrieved from: http://www.nato.int/cps/en/natohq/topics_110496.htm. Accessed 5 Nov 2016
NATO (2001) Statement by NATO secretary general, Lord Robertson, 10 Feb 2001. Retrieved from: http://www.nato.int/docu/speech/2001/s011002a.htm. Accessed 25 Nov 2016
Nguyen Reese (2013) Navigating Jus ad Bellum in the age of cyber warfare. California Law Review 101:1079
O'Connell Mary Ellen (2002) Lawful self-defense to terrorism. University of Pittsburgh Law Review 63:889
OAS (2001) Terrorist threat to the Americas. OAS Ministerial Resolution, September 21, 2001, RC.24/RES.1/01
Schmitt M (2013) Tallinn manual on the international law applicable to cyber warfare. Cambridge University Press, Newyork
UN (1945) Charter of the United Nations, 24 Oct 1945, 1 UNTS XVI. Retrieved from: http://www.refworld.org/docid/3ae6b3930.html. Accessed 29 Oct 2016
UN (2001) Draft articles on the responsibility of states for internationally wrongful acts, Report of the ILC on the work of its fifty-third session, UN GAOR, 56th Sess, Supp No 10, 43, UN Doc A/56/10
UNSC (2001) Security Council resolution 1368 (Threats to international peace and security caused by terrorist acts), 12 September 2001, S/RES/1368. Retrieved from: http://www.refworld.org/docid/3c4e94557.html. Accessed 13 Oct 2016
US Code (2001) 42 US Code §5195c—critical infrastructures protection. Retrieved from: https://www.law.cornell.edu/uscode/text/42/5195c. Accessed 23 Nov 2016
Washington Post (2014) Sony pictures hack appears to be linked to North Korea, investigators say, 12/03/2014. Retrieved from: https://www.washingtonpost.com/world/national-security/hack-at-sony-pictures-appears-linked-to-north-korea/2014/12/03/6c3c7e3e-7b25-11e4-b821-503cc7efed9e_story.html. Accessed 19 Oct 2016
Wilmshurst E (2005) Principles of international law on the use of force by States in self-defense. Royal Institute of International Affairs. Working Paper

Author Biography

Sonia Boulos obtained an LL.M. and a JSD in international human rights law from the University of Notre Dame, USA. She currently works as an adjunct professor at the University of Antonio de Nebrija in Madrid. Her research interests include international human rights law, international law and the use of force and the doctrine of Responsibility to Protect. In addition, Boulos acquired an extensive experience working as a staff attorney with leading Israeli and Palestinian human rights organization.

War-Like Activities in the Cyberspace: Applicability of the Law of Armed Conflicts

Jerónimo Domínguez-Bascoy and Bartolomé Bauzá-Abril

Abstract The security of cyberspace high strategic interest has been particularly intensified since the States have become aware of the possibility of extending their military operations to that virtual space, which has thus become a "fifth domain of warfare", adding to the terrestrial, maritime, air and space domains. This raises the question of whether current conventional and customary International Humanitarian Law applies to cyber-hostilities, that is, the military operations that the parties to an armed conflict conduct against the adversaries in and through cyberspace. Although the analysis of the application of the current *jus in bello* to the cyber operations obviously includes many other aspects, the present chapter addresses only two basic issues. The first one focuses on the requirements that, in the absence of conventional hostilities, should concur for the applicability of International Humanitarian Law to operations conducted by cyber means. This makes it necessary to differentiate between international and non-international armed conflicts. The second one is about when cyber operations would be equivalent to "attacks" in the sense of the *jus in bello*: acts of violence against the adversary, whether in offence or in defence, paying particular attention to the "violence" element of the definition.

Keywords Cyberspace · Cyberwarfare · Armed conflicts · War · IHL

Submitted: 26.11.16; Accepted: 20.12.16.

J. Domínguez-Bascoy (✉)
Spanish Armed Forces Legal Service, Spanish Navy General Headquarters, Madrid, Spain
e-mail: jrodbas@fn.mde.es

B. Bauzá-Abril
Outrospection SL, Madrid, Spain
e-mail: bbauza1@gmail.com

J.M. Ramírez and L.A. García-Segura (eds.), *Cyberspace*, Advanced Sciences and Technologies for Security Applications, DOI 10.1007/978-3-319-54975-0_15

1 Cyberspace as a New Battleground

War-like activities in cyberspace, or cyber-warfare as it is commonly known, are becoming more and more relevant for the international community in many aspects, the legal one included. Some activities in cyberspace result in potential risks and threats to this global and dynamic domain made up of information and communication infrastructures, networks and systems. Security in cyberspace has already become a strategic issue as States have begun to understand the advantages of protecting, and even extending their military activities, to this virtual battleground, which has been considered as the fifth domain of warfare, the other domains being Land, Maritime, Air and Space. Some States are then creating cyber-specialized units in their armed forces, providing the knowledge and technical means to conduct war-like activities in cyberspace.

It is not then surprising that issues as the applicability of current conventional and customary International Humanitarian Law (IHL) to cyber-hostilities, that is, to military operations in and through cyberspace that parties engaged in an armed conflict direct against an adversary,[1] had been raised for some years now.

One of the first official statements in that regard is contained in the *International Strategy for Cyberspace. Prosperity, Security, and Openness in a Networked World*, published by the Obama administration in 2011. There, it is said that

> The development of norms for state conduct in cyberspace does not require a reinvention of customary international law, nor does it render existing international norms obsolete. Long-standing international norms guiding state behavior—in times of peace and conflict—also apply in cyberspace (White House 2011).

Shortly after, in September 2012, during a meeting hosted by the United States Cyber Command (USCYBERCOM), the then Legal Advisor to the US State Department, Harold Koh, advanced this same idea, by proclaiming the applicability of the principles of international law in cyberspace. He insisted in that cyberspace is not a law-immune area, open to unregulated, unrestricted hostile activities, and that —as in any other context of armed conflict—the use of force, even in self-defence, is limited by the principles of necessity and proportionality (Schmitt 2012).

In September 2014, at the Cardiff Summit, NATO explicitly recognized "that international law, including international humanitarian law and the UN Charter, applies in cyberspace" (NATO 2014). More recently, in a communiqué issued after the Warsaw Summit, in 2016, after recognising that cyberspace is "a domain of operations in which NATO must defend itself as effectively as it does in the air, on land, and at sea", NATO reaffirms the commitment to act in accordance with international law, including the UN Charter, international humanitarian law, and human rights law, as applicable (NATO 2016).

[1]These are operations directed against or launched from a computer or a system with the objective of infiltrating a system to retrieve, extract, destroy, change or encrypt data up to the manipulation of an industrial infrastructure to control its performance.

As for the United Nations, the last landmark in the process officially referred to as "*Developments in the field of information and telecommunications in the context of international security*", started back in 1998, is a report issued by a Group of Governmental Experts, the fourth of its kind, in June 2015. One of the key aspects in this report is on the applicability of international law to the use of information and communications technologies (ICT). The experts start by recalling the statement contained in the previous Group of Governmental Experts' report (2013), where it was said that international law, particularly the UN Charter, is applicable and is essential to maintaining peace and stability and promoting an open, secure, peaceful and accessible ICT environment. As regards, more specifically, to the way International Humanitarian Law applies to the use of the ICT by states, the only statement made is that there are established international legal principles, including, where applicable, the principles of humanity, necessity, proportionality and distinction [UN General Assembly (2015)].

Once it is accepted the applicability of international law rules and regulations to cyberspace, especially to aspects related to *jus ad bellum* and *jus in bello*, it is necessary to grasp how these rules and regulations can be shaped into the specificities of cyberspace. So far, one of the most remarkable efforts is the *Tallinn Manual on the International Law Applicable to Cyber Warfare.* This Manual presents the results of the discussions held by a group of experts, between 2009 and 2012, led by Professor Michael N. Schmitt. Topics in *The Tallinn Manual* include State sovereignty and responsibility in cyberspace, *jus ad bellum*, International Humanitarian Law, and Law of Neutrality. It is structured in chapters and rules within. Each one of these rules is the result of the unanimous acceptance of the experts, and is supported by extensive discussion on existing treaties and customary law and the interpretations the experts make on their applicability to cyberspace. Disagreements within the group as to the application of each rule are also included.

2 Preconditions for the Application of IHL to Cyber Hostilities

Massive use by the media of expressions such as cyberattack, or even cyberwar, could lead us to the idea that we are witness of hostilities in cyberspace. The truth is that for the IHL to be applicable to the cyber operations thus labelled, they would have to take place in the context of an armed conflict. As we know them, in most cyber operations, such a context is missing, non-existent. Therefore, instead of cyber hostilities, we should be referring to cybercrime, cyber vandalism, cyber espionage or cyber terrorism.

It is commonly agreed that when cyber operations are conducted in the context of an ongoing armed conflict, where conventional means and methods of warfare are being also used, those cyber-operations will be subject to the rules of IHL as they are applicable to the conflict, be it international or non-international. What it is under discussion is when cyber-operations, by themselves, can trigger the application of IHL. To answer it, precise requisites must be considered to ascertain if we are in presence of one of two types in which, despite the new factual situations of the current reality, the ICRC considers it is possible to continue dividing the armed conflicts.

2.1 *Cyber Operations and International Armed Conflicts*

The Geneva Conventions declares that they:

> …shall apply to all cases of declared war or of any other armed conflict which may arise between two or more of the High Contracting Parties, even if the state of war is not recognized by one of them (Common Article 2).

The generally agreed definition of international armed conflict is the one formulated by the United Nations International Criminal Tribunal for the former Yugoslavia in the Tadic case (1997), where it was considered that such a conflict occurs "whenever there is a resort to armed force between States".

Therefore, to determine the conditions which should be met for cyber-operations, for themselves, to unleash an international armed conflict what it must firstly be assessed if such cyber-operations can be attributed to a State, and, secondly, to what extent they could be equated with that "*resort to armed force*" mentioned in the Tadic judgment.

Attribution to a State

Quite rightly, Droege (2012) writes that attribution in cyberspace is particularly difficult as anonymity is more a rule than an exception. In view of the possibility of resorting to legal presumptions, and bearing in mind the rule expressed by the International Court of Justice in 1949 in the Corfu Channel case-, according to which no State can knowingly allow the use of its own territory to carry out acts contrary to the lawful rights of any other State, Droege radically rejects the uninformed presumption of State attribution from the mere fact that a cyberattack has being launched from the governmental infrastructure of a State. Droege notes that such presumption would be quite excessive in a context of cyberwarfare, as IT systems are difficult to protect from manipulations and the relative easiness to remotely control them. Again, this overstated presumption, continues DROEGE, would place a very heavy burden on States, as they would be held responsible of the full spread of home-based IT operations, without additional evidences. In this sense, Rule 7 of the Tallinn Manual states that

> the mere fact that a cyber operation has been launched or otherwise originates from governmental cyber infrastructure is not sufficient evidence for attributing the operation to that State, but is an indication that the State in question is associated with the operation.

An additional issue, highlighted by the cyber operations directed against Estonia in 2007 and Georgia in 2008, is related to State attribution of cyber-operations carried out by non-state actors. Melzer (2011) argues that the applicability of IHL cannot be limited to acts committed by members of the State military but must be extended to the all personnel acting *de jure* or de facto, as a State agent.

The *Draft articles on Responsibility of States for Internationally Wrongful Acts*, drafted by the International Law Commission, states that

> the conduct of a person or group of persons shall be considered an act of a State under international law if the person or group of persons is in fact acting on the instructions of, or under the direction or control of, that State in carrying out the conduct (UN 2001).

Regarding the intensity of this direction or control, two standards have already been set up by the international jurisprudence. The first one was established by the International Court of Justice in the case concerning military and paramilitary activities in and against Nicaragua brought by Nicaragua against the United States (1986), according to which it would be required a State effective control over each and every one of the cyber operations carried out by non-State actors. The other one was adopted by the Appeals Chamber of the International Criminal Tribunal for the Former Yugoslavia in the Tadic case, according to which an overall State control over the activities of such non-state actors would suffice, as long as they would be part of an organized group. Thus, an effective control over any specific cyber-operation would not be required.

The Tallinn Manual, in its commentary to rule 22, contends that mere State support for a group of non-State actors involved in a non-international armed conflict will not be enough to *internationalise* the conflict. Consequently, there was a general agreement within the group of experts that the threshold for internationalization is a high one: merely taking measures to maintain rebel access to the national cyberinfrastructure was not considered by the Experts to suffice. Similarly, the provision of cyber-attack tools for rebel use would not reach the threshold. By contrast, providing specific intelligence on cyber vulnerabilities that renders rebel cyber-attacks possible would, in their view, suffice.

Resort to Armed Force

Droege (2012) argues that when conventional means and methods of warfare, such as aerial bombardments, artillery fires o troop deployments, are used, there is usually no controversy on the meaning of such actions as a resort to armed force. Additionally, Schmitt (2014) puts forward the idea that from the *jus in bello* perspective the concept of an armed force requires some form of hostilities: in other words, there must exist a "collective resort by the parties to the conflict to means and methods of injuring the enemy" (ICRC 2009).

Furthermore, in the ICRC *Commentary to the Geneva Conventions* of 1949 it is said that:

> any difference arising between two States and leading to the intervention of armed forces is an armed conflict within the meaning of Article 2, even if one of the Parties denies the existence of a state of war. It makes no difference how long the conflict lasts, or how much slaughter takes place (Pictet 1952, p. 32).

Translating these concepts to the realm of cyberspace is not an easy task. As cyberspace is immune to the kinetics reigning over the conventional combat means and methods, an additional effort should be made to determine what activities in this virtual domain would amount to a use of armed force.

Perhaps, the consequences of a specific cyber-operation can be considered as a starting point, verifying if an analogy can be established with the consequences of the use of conventional means and methods of warfare. This will be the case when, for example, a cyber-attack on a SCADA system of a critical infrastructure, such as an air traffic control system (ATC), results in damage to property or injury to persons.

Nevertheless, Droege emphasizes that the abovementioned criterion does not in itself fully capture all the effects associated to cyber-operations and the damages they may cause, which do not necessarily resemble those of traditional weapons. Most probably, she continues, the resort to cyber-operations will not be intended to destroy or physically damage a military or civil infrastructure, but to impede its normal operation. As an example, a cyber-attack could result in the collapse of an electrical grid, with no physical damage inflicted. In such cases, even in the absence of conventional military means or immediate physical destruction, the potential impact of these cyber-activities upon the civilian population could be more severe or intense than the physical destruction of a block of buildings and would support considering such cyber-operations as a resort to armed force.

Some experts emphasize not just the similarity of the effects a cyber-operation may have compared to conventional operations, but the combination of factors such as severity of the results of the cyber-operation, the assets employed, the involvement of the military or other governmental sectors, the nature of the target—civilian or military—or the duration of the operation in question.

To illustrate the difficulty of determining when a cyber-operation would constitute an armed case, the Tallinn Manual addresses the 2010 Stuxnet operation against SCADA systems in Iran, as a result of which centrifuges at a nuclear fuel processing plant, in Natanz (Iran) were physically damaged, action attributed by the New York Times to the joint action of The United States and Israel (see for instance, Langer 2013). The International Group of Experts was divided as to whether the damage sufficed to meet the armed force criterion. This case exemplifies some of the complexities involved in determining if or when a cyber-operation acts as a trigger for an international armed conflict. We must wait for the development of some State practice and an *opinio iuris* in this area.

2.2 Cyber Operations and Non-International Armed Conflicts

The Tallinn Manual points out that:

> a non-international armed conflict exists whenever there is protracted armed violence, which may include or be limited to cyber operations, occurring between governmental armed forces and the forces of one or more armed groups, or between such groups", and "the confrontation must reach a minimum level of intensity and the parties involved in the conflict must show a minimum degree of organization (Rule 23).

It is, then, a definition derived from Common Article 3 to the Geneva Conventions and the jurisprudence of the International Criminal Tribunal for the Former Yugoslavia in the Tadic case on the basis of which it is widely accepted that there are two basic criteria to determine when we are facing a non-international armed conflict: the intensity of hostilities and the involvement of an organized armed group. It is necessary, then, to understand what problems pose the application of these criteria to cyber-operations.

Intensity

Unlike international armed conflicts, non-international ones require the level of violence to reach a certain threshold. Following the Additional Protocol II to the Geneva Conventions, it is then generally accepted that "situations of internal disturbances and tensions, such as riots, isolated and sporadic acts of violence and other acts of a similar nature" do not reach the level of violence required to be considered armed conflicts. From there, the commentary to Rule 23 in the Tallinn Manual argues that sporadic cyber-incidents, including those directly causing physical damage to people or materiel will only in exceptional cases amount to a non-international armed conflict.

Doctrine and international jurisprudence refer to several factors to assess the intensity of hostilities in order to ascertain whether we are dealing with a non-international armed conflict. Thus, the resort to military force and not just constabulary force to confront violence; the type of weaponry used; the territorial extension of hostilities; the number of civilians forced to flee the combat zones; or the amount of physical damage and the number of casualties produced.

As Droege points out, cyber-operations, considered in themselves, would not exhibit many of the previous factors to evaluate the intensity of violence, but—she says—their consequences could, however, reach the required level of violence, as, for example, would happen when as a result of a cyber-attack the floodgates of a reservoir would be open; or mid-air collisions be prompted.

Melzer argues that to the extent that cyber-operations are being launched from the target State's own territory and are not supported by the threat or use of military conventional forces, such as preventing the target State the exercise of its territorial authority upon the attackers, they will most likely be understood as criminal activities that should be dealt via constabulary actions rather than through the armed conflict paradigm. Conversely, most likely, such cyber-operations will be qualified as *hostilities* liable to unleash a non-international armed conflict when they are recurrent—not sporadic-, originate from a territory over which the target State cannot exercise coercive activities and, at the same time, local authorities are either unable or unwilling to act to stop those cyber-operations.

At any rate, as Schmitt claims, there is no reason at all to suggest that the intensity criterion as applicable to non-international armed conflicts is going to be weakened, as would not be rational to sustain that the use of cyber-assets and means should enjoy a lower threshold compared to conventional kinetic operations.

Organization of the Armed Group

ICRC's report on "*International Humanitarian Law and the challenges of contemporary armed conflicts*" (2011, Note 10) contends that international jurisprudence has identified some gauges to ascertain if the *organization* requisite exists in the non-State part of a non-international armed conflict. They include the existence of a command structure and disciplinary rules and mechanisms within the armed group, the existence of headquarters, the ability to procure, transport and distribute arms, the group's ability to plan, coordinate and carry out military operations, including troop movements and logistics, its ability to negotiate and conclude agreements such as cease-fire or peace accords, etc.

In the commentary to Tallinn's rule 23, the question of *virtual* organization in which all activities that bear on the criterion occur on-line is raised. It is then argued that the fact of a number of hackers are simultaneously attacking a State is not by itself an indication of an organization. Quite a different thing would be a distinct online group with a leadership structure that coordinates its activities by, for instance, allocating specified cyber targets amongst themselves, sharing attack tools, conducting cyber vulnerability assessments, and doing cyber damage assessment to determine whether 'reattack' is required. In this case, we would be dealing with a *cooperative* action in which, according to the majority of the International Group of Experts the fact that the failure of members of the group physically to meet does not alone preclude it from having the requisite degree of organisation.

Droege sees quite improbable that groups of hackers or groups virtually communicating or linked may have the organization or command structure—including the ability to take disciplinary measures—required to be considered part of any conflict.

3 Basic Issues Related to the Application of IHL to Cyber Hostilities

3.1 Determination of the Cyber Operations Subject to IHL Rules on the Conduct of Hostilities

The fundamental norm in matters related to conduction of hostilities is the Additional Protocol I to the Geneva Conventions. There, it is specified one of the cardinal principles in IHL, the principle of distinction, according to which

> In order to ensure respect for and protection of the civilian population and civilian objects, the Parties to the conflict shall at all times distinguish between the civilian population and combatants and between civilian objects and military objectives and accordingly shall direct their operations only against military objectives (Article 48).

Although the above formulation refers broadly to *operations*, truth is that the subsequent articles ruling the conduction of hostilities refer specifically to *attacks*. Then, the argument goes, it has to be determined whether the rules related to conduction of hostilities apply exclusively to attacks or, perhaps more appropriately, to the wider spectrum of military operations.

Schmitt argues that is in this series of proscriptions and restrictions of attacks where Additional Protocol I operationalizes article 48. For example, when states that:

> The civilian population as such, as well as individual civilians, shall not be the object of attack (Article 51.2); Indiscriminate attacks are prohibited (Article 51.4); Civilian objects shall not be the object of attack or of reprisals (Article 52.1); It is prohibited to attack, destroy, remove or render useless objects indispensable to the survival of the civilian population (Article 54.2); Attacks against the natural environment by way of reprisals are prohibited (Article 55.2); etc. Consequently, only those cyber-operations meeting such requirements as identified by IHL—requirements that will be addressed below—are to be regulated in accordance with rules dealing with the conduction of hostilities (Schmitt 2015a).

Quite a different view is held by Melzer, who sustains that however much the term *attack* is a key notion in IHL, any analysis on the relevance of the Articles on conduction of hostilities and their application to cyber-hostilities, cannot be limited to this notion. He goes on saying that the basic rule on *distinction*—as in the Additional Protocol I—refers to *operations* and not to *attacks*. And he mentions other articles that refer as well to *operations* as is, for example, Article 51.1 that asserts that "the civilian population and individual civilians shall enjoy general protection against dangers arising from military operations" (Melzer 2011). For him, although attacks certainly represent the predominant form of combat operations, it would be inaccurate to assume that all cyber-operations short of attack are to be exempt of the IHL regulations on conduction of hostilities. Therefore, in his view, the application to cyber-operations of the restrictions imposed by IHL on the conduct of hostilities will not depend on whether the operations in question can be described as *attacks*, but whether they constitute an integral part of the *hostilities*, with the meaning that this term has in IHL.

Droege, for her part, believes that this debate is purely formalist, since the fundamental question underlying it should, more properly, be tackled around the concept of *attack* and, in particular, how should it be understood taking into account the very peculiar characteristics that, compared to the kinetic ones, exhibit the operations in the cyberspace. This will be addressed next.

3.2 Cyber Operations as "Attacks"

Additional Protocol I, Article 49.1, affirms that "attacks means acts of violence against the adversary, whether in offence or in defence". Such definition serves as a basis to the definition of cyber-attack formulated in Rule 30 of the Tallinn Manual: "a cyber-operation, whether offensive or defensive, that is reasonably expected to cause injury or death to persons or damage or destruction to objects". Although Article 49.1 of the Protocol Additional I refers to attacks as *acts of violence*, there is a general consensus on that what is relevant is not the violence of the act in itself, but the violent consequences derived from it. Specifically, such injuries, death, damage or destruction that are traditionally associated to attacks with conventional weapons, even though some of them, as is the case with biological, chemical or radiological weapons… and cyber-weapons, do not use kinetic force.

Consensus disappears and controversy takes its place at the time of determining if—because of its specific nature—cyber-operations equivalent to attacks (as per Additional Protocol I) would be strictly reduced to those whose effects were injuries, death, damage or destruction; or rather the concept of cyber-attack had to be expanded to include such operations as to cause serious effects, other than those listed above. Some years ago, Dörmann (2004) supported this second option by sustaining that cyber-operations can also be considered attacks even though damage or destruction of an object is not derived from them. To support his view, he observes the definition of military objective as formulated by Article 52.2 of the Additional Protocol I, that includes those "whose total or partial destruction, capture or neutralization, in the circumstances ruling at the time, offers a definite military advantage. This reference to neutralization" in defining attacks whose aim is to negate the use of an object to the enemy, without physically destroying it, is precisely what Dörmann uses to sustain that the way through which an object is rendered unusable is quite irrelevant.

Schmitt makes evident that a cyber-operation with no physical effects intended directed against a civilian cyber-infrastructure can, in some cases, be much more harmful than another conventional operation aiming at causing limited damage. Droege provides some useful examples: disrupting the civilian electrical grid or water treatment system; directing a denial of service attack on an Internet banking system with significant impact on the ability of a few million bank customers to access banking services; directing a denial of service attack on a private airline's online booking system in order to cause inconvenience to the civilian population; blocking access to Facebook for the entire population because it contains pro insurgency propaganda; etc.

The need to determine when such cyber-operations would qualify as attacks has led to propose two basic doctrinal criteria synthesized by Schmitt. According to the first, all *data* should have the same consideration as objects have, so that any cyber-operation intended to manipulate, alter or delete civilian data would be prohibited. Nevertheless, the author himself has rebuked this approach (Schmitt 2015b) on the grounds of the *tangibility* of the object, which data lack. Additionally, the number of prohibited cyber-operations (for example, deleting a civilian blog or a post from a blog in the Internet would constitute a violation of IHL) during an armed conflict would become inconvenient, significantly altering that balance between humanitarian concern and the military need that has always sought to safeguard the IHL.

The second criterion, which was maintained by most of the experts involved in the preparation of the Tallinn Manual, leads to the inclusion of *loss of functionality* within the concept of harm. The logic of this approach, shared by Droege, is that *damage* is a different concept from *destruction*, consequently, disruption of systems by interfering in the computing systems upon which they depend on is equivalent to *damage*, as there is a loss of utility. As a note of caution, she continues that by including the interferences on civilian systems into this concept, we could also be incurring in a dangerous broadening of the term *attack*. In her view, considering that the disruption of civilian communications by electronic messaging systems or social networks fall into the category of *attack* goes far beyond of it was intended when regulating the conduction of hostilities. Accordingly, Droege explores the object and intent of current IHL norms to identify which cyber-operations interfering with the functionalities of certain systems would be excluded from the concept of *attack*. Consequently, cyber-operations that are equivalent to espionage, to the dissemination of propaganda, to embargoes, or other non-physical means of psychological or economic warfare will not fall under the definition of *attacks*. Also, the term *attack* cannot comprise cyber operations that would be tantamount to economic sanctions, unless they would have an impact on objects affecting to the survivability of the civilian population. Finally, the conduct of cyber-operations that interfere with civilian communication systems, whose parallel in the physical world is probably the jamming of radio communications or television broadcasts, should be excluded.

In short, without prejudice to further State practice, we believe that when a cyber-operation is not merely limited to cause an effect on computer systems but goes beyond that and intends to produce effects in the physical world by disrupting the functioning of computer-controlled infrastructures, such as an electricity distribution network, an air traffic control system or a banking services, what we are witnessing are acts constituting an attack in accordance with *jus in bello*.

4 Conclusion

The high strategic value that the use of cyberspace has in itself, given the great dependence our societies have from it on virtually all the spheres of their functioning, has been increased by the fact that this virtual space can also be used as a new avenue for the use of armed force.

The possibility that, together with traditional means and methods of warfare, the conduct of hostilities is accompanied by the use of cyber means and methods of warfare, raises the inevitable question as to whether the current rules of jus in bello are fully applicable to those operations conducted in and through cyberspace.

A number of Western states, and some international organizations such as NATO, have already expressed their views on the matter in the affirmative. NATO itself, through its Cooperative Cyber Defense Center of Excellence, has sponsored the largest doctrinal effort ever undertaken to determine how the IHL should be applied to hostilities conducted in cyberspace: the "Tallinn Manual on the International Law Applicable to Cyber Warfare".

However, the non-kinetic nature of cyber operations leaves many doubts in the air, which can only be resolved through the practice of States or, where appropriate, through the agreements reached in intergovernmental forums, such as the one that is taking place within the United Nations under the name of "Developments in the field of information and telecommunications in the context of international security".

Although wars conducted exclusively through cyber means and methods of warfare do not appear to be very likely, there are many International Law experts who have attempted to translate into cyberspace the categories used in determining the precise requirements for an armed conflict, whether international or non-international, to which the IHL is applicable.

Likewise, there has been a lively debate about to what kind of cyber operations the norms of jus in bello that govern the conduct of hostilities would apply. In particular, it has been debated on what type of cyber operations would be equivalent to "attacks", as this term is conceived in jus in bello. However, there seems to be an agreement that, when making such equivalence, attention must be directed to the violent consequences of cyber operations. Consequences that, as has been pointed out, can have different scopes. Thus, Roscini has distinguished three types of effects of cyber operations: primary, that would be those produced in the data and the software object of the attack; secondary, those produced on the infrastructures controlled by the attacked system; and tertiary, which would be those produced on the people affected by the destruction or incapacitation of the system or infrastructure attacked (Roscini 2014).

References

Dörmann K (2004) Applicability of the additional protocols to computer network attacks ICRC. Retrieved from: https://www.icrc.org/eng/assets/files/other/applicabilityofihltocna.pdf
Droege C (2012) Get off my cloud: cyber warfare, international humanitarian law, and the protection of civilians. International Review of the Red Cross 94(886):533–578. Retrieved from: https://www.icrc.org/eng/assets/files/review/2012/irrc-886-droege.pdf
International Committee of the Red Cross (2009) Interpretive guidance on the notion of direct participation in hostilities under international humanitarian law ICRC. Retrieved from: https://www.icrc.org/eng/assets/files/other/icrc-002-0990.pdf
International Committee of the Red Cross (2011) The notion and typology of armed conflicts in international humanitarian law and the challenges of contemporary armed conflicts, 31st international red cross conference, Geneva. Retrieved from: https://app.icrc.org/e-briefing/new-tech-modern-battlefield/media/documents/4-international-humanitarian-law-and-the-challenges-of-contemporary-armed-conflicts.pdf
Langer R (2013) To kill a Centrifuge. Retrieved from: http://www.langner.com/en/wp-content/uploads/2013/11/To-kill-a-centrifuge.pdf
Melzer N (2011) Cyberwarfare and international law. United Nations Institute for Disarmament Research (UNIDIR). Retrieved from: http://unidir.org/files/publications/pdfs/cyberwarfare-and-international-law-382.pdf
NATO (2014) Wales summit declaration. Cardiff, 5 Sept 2014. Retrieved from: http://www.nato.int/cps/en/natohq/official_texts_112964.htm
NATO (2016) Warsaw summit declaration. Warsaw, 9 July 2016. Retrieved from: http://www.nato.int/cps/en/natohq/official_texts_133169.htm
Pictet JS (1952) Commentary to the Geneva convention for the amelioration of the condition of the wounded and sick in armed forces in the field. ICRC, Geneva
Roscini M (2014) Cyber operations and the use of force in international law. Global Oup. https://global.oup.com/academic/product/cyber-operations-and-the-use-of-force-in-international-law-9780199655014?cc=es&lang=en&
Schmitt MN (2012) International law in cyberspace: the Koh speech and Tallinn manual juxtaposed. Harvard Int Law J 54. Retrieved from: http://www.harvardilj.org/wp-content/uploads/2012/12/HILJ-Online_54_Schmitt.pdf
Tallinn Manual on the International Law Applicable to CyberWarfare (2013). Cambridge University Press
Schmitt MN (2014) The law of cyber warfare: Quo Vadis? Stanford Law Policy Rev 25:269–299. Retrieved from: https://journals.law.stanford.edu/sites/default/files/stanford-law-policy-review/print/2014/06/schmitt_25_stan._l._poly_rev._269_final.pdf
Schmitt MN (2015a) The law of cyber targeting. Tallinn Paper No 7, NATO CCDCOE. Retrieved from: https://ccdcoe.org/sites/default/files/multimedia/pdf/TP_07_2015.pdf
Schmitt MN (2015b) The notion of 'objects' during cyber operations: a risposte in defence of interpretive and applicative precision. Israel Law Review, pp 81–109. Retrieved from: http://papers.ssrn.com/sol3/papers.cfm?abstract_id=2557989
UN General Assembly (2015) Group of governmental experts on developments in the field of information and telecommunications in the context of international security. New York, 22 July 2015. Retrieved from: http://www.un.org/ga/search/view_doc.asp?symbol=A/70/174
UN International Law Commission (2001) Draft articles on responsibility of states for internationally wrongful acts, with commentaries. Document A/RES/56/83. Yearbook of the International Law Commission II, Part Two. Retrieved from: http://legal.un.org/ilc/texts/instruments/english/commentaries/9_6_2001.pdf

UN International Tribunal for the Prosecution of Persons Responsible for Serious Violations of International Humanitarian Law in the Territory of the former Yugoslavia (1997) Case nr. IT-94-1-T, Prosecutor v. Tadic, 7 May 1997. Retrieved from: http://www.icty.org/x/cases/tadic/tjug/en/tad-tsj70507JT2-e.pdf

USA White House (2011) International strategy for cyberspace. Prosperity, security, and openness in a networked world. Retrieved from: https://www.whitehouse.gov/sites/default/files/rss_viewer/international_strategy_for_cyberspace.pdf

Author Biographies

Col. Jerónimo Domínguez-Bascoy joined the Spanish Navy Legal Service in 1982. He holds a Diploma in Social Communication from the School of Information Sciences, Universidad Complutense de Madrid (UCM), and a Diploma in International Military Law. In 2013, he completed a Master in Security and Defence at the UCM. He has served at the Navy's Legal Office in the Canary Islands, Las Palmas de Gran Canaria, and in many other military billets. Col. Domínguez Bascoy was Legal Advisor to NATO's Joint Sub-Regional Command South West Headquarters (2002–2004); to the European Union Naval Force Headquarters in the Indian Ocean, in Operation ATALANTA against piracy (2011); and, as an embarked staff member, in several NATO Response Force exercises. He completed the Advanced Course on the Law of Armed Conflict at the International Institute of Humanitarian Law, Sanremo (Italy). At present, he is posted in the SP Navy Staff HQ Legal Office in Madrid.

Rear Admiral Bartolomé Bauzá-Abril retired from the Navy in 2014, after a 40-year career, a good part of it on board surface ships. His last assignment was at Navy HQ, Madrid, where he was in charge of the Communications and Information Systems Division. Amongst other tasks, he was responsible for the initial setup of the Spanish Navy Cyberdefense capability. This was a challenge covering a myriad of aspects, one of them understanding the legal characterization of the cyberspace. He has also been DCOM of European Union "Operation Atlanta". After retiring, he founded Outrospection SL, a company dealing with Business Development Support, Leadership, Security and Transparency.

Negotiation on Cyber Warfare

Carlo Trezza

Abstract In spite of the growing the risk that cyber capacities could be used as military tools no international agreement has been finalized so far to prevent a cyberwarfare involving sovereign states as well as non-state actors. States have different conceptual approaches and priorities when approaching this subject and discussing definitions. There are additional peculiarities. The possession and use of cyber weapons capacities are not visible. The author of an attack cannot be clearly identified, and non-state actors can play at the same level as national states. The operators of such weapons can hardly be included in the classical legal category of "combatant". Most countries have already established cyber structures integrated into their military chains of command and fully dedicated them to cyber defense and cyber offense. The cases of the US, Nato, European Union and the United Nations. Cyberwarfare has not yet acquired a legal status of its own in spite of the fact that after land, sea, air and outer space, cyberspace has become the fifth domain in which states can confront each other militarily. States have not even started to negotiate any sort international regulation. The Tallinn Manual on the International Law applicable to cyber warfare has no international legal value but has the merit of having established some fundamental principles. It indicates that the norms applicable to cyber are the same as those applicable to the other types of weapons and in particular the International humanitarian law. Nothing prevents the international community from considering cyberwarfare also in a preventive mode, and to craft cyber-specific rules prohibiting cyber instruments capable of provoking catastrophic consequences. A general prohibition of possession and use of cyber offensive capabilities would ideally be the preferable solution but we are nowhere close to such a solution. An uncontrolled cyber incident could act as a shock

Outgoing Chair of the Missile Technology Control Regime (MTCR).

Submitted: 9.8.16; Accepted: 13.9.16.

C. Trezza (✉)
United Nations Office for Disarmament Affairs (UNODA), New York, USA
e-mail: carlotrezza@icloud.com

J.M. Ramírez and L.A. García-Segura (eds.), *Cyberspace*, Advanced Sciences and Technologies for Security Applications, DOI 10.1007/978-3-319-54975-0_16

absorber to prevent a conflict from escalating; it could also become a trigger for a wider confrontation that the international community cannot risk. The problem of how to address the security implications of a cyber world will be with us for the years to come. Better to address the issue in a preventive mode rather than subsequently to a possible cyber confrontation.

Keywords Cyberwarfare · Weapons of mass destruction · The Shanghai Cooperation Organization · USCYBERCOM · NATO strategic concept

1 Introduction

A cyberworld has existed for more than three decades; in parallel a military/security dimension has also been developed. It could possibly lead to a confrontation between states and to a cyber "arms race". In spite of this clear and present danger and of the risk that cyber capacities could be used by terrorist groups, no international agreement has been finalized so far to prevent a cyberwarfare involving sovereign states as well as non state actors. In his foreword to the latest UN report on cyber activities the UN Secretary-General Ban Ki Moon noted that

> the benefits (of cyberspace) are enormous, but these do not come without risk. Making cyberspace stable and secure can only be achieved through international cooperation, and the foundation of this cooperation must be international law and the principles of the UN Charter (Ban Ki Moon, UN General Assembly 2015a, b, c).

The UNSG was correct in indicating that the existing international norms must be the basis to achieve stability and security of cyberspace. Implementing laws already in existence is necessary, but it may not be sufficient to attain those goals: additional rules, better tailored to the military/security cyber environment and to its peculiarities are to be pursued.

2 To Which Category of Weapons Cyber Weapons Belong?

It is difficult to establish the category to which cyber weapons belong. Are they to be considered conventional weapons or weapons of mass destruction (WMD)? The distinction between these two categories, although artificial, is fundamental: only nuclear, chemical and biological are considered as WMDs. Cyber, by exclusion, would therefore fall into the category of conventional weapons. This has legal consequences. According to UN Security Council Resolution 1540, which is a legally binding resolution under Chapter Vll of the UN Charter, only the proliferation od WMDs is a threat to international peace and security:

> Affirming that proliferation of nuclear, chemical and biological weapons, as well as their means of delivery, constitutes a threat to international peace and security (UNSC 2004, first paragraph).

Thus proliferation of cyber weapons and cyber technology, at least from a legal stand, does not represent a threat to international peace and security. Their possession is not prohibited and their use and threat of use are only subject to the general international rules on warfare.

3 Conceptual Approaches and Priorities

The issue of cyberwar is further complicated by the fact that states have different conceptual approaches and priorities when discussing definitions. The six Eastern countries members of the Shanghai Cooperation Organization (SCO)—People's Republic of China, Kazakhstan, Kyrgyz Republic, Russia, Tajikistan, and Uzbekistan—defined cyberwar to include dissemination of information "harmful to the spiritual, moral and cultural spheres of other states" (2008, unofficial translation). What they fear is that freedom of information and the penetration of political or religious ideas through the cyber networks might destabilize national societies and become a mortal threat to the survival of their regimes. This is not a farfetched hypothesis. Let us only consider the role played by electronic devices in the hands of demonstrators during the so called "Arab Spring".

Western countries have different priorities. They view cyberwar as causing effective physical damage and injury to their military assets and strategic infrastructures. They are more open to freedom of speech and information and in principle they oppose limitations to the circulation of political and religious ideas via the cyber network. And yet episodes in the West of whistleblowers being prosecuted, investigation agencies asking cyber companies to reveal their codes, indicate that even Western countries feel vulnerable to a total liberalization.

There are additional peculiarities. The possession and use of cyber weapons capacities is not visible. The author of an attack cannot be clearly identified and non-state actors can play at the same level as national states. The tradition of starting a military confrontation through an official war declaration is no longer applicable and its is more and more difficult today to establish if and when a war situation starts. The operators of such weapons can hardly be included the classical legal category of "combatant"; they rather belong to a new profile that can be defined as "armchair warriors." As the pilots of unmanned aerial vehicles, better known as UAVs, the cyber operators use their instruments of war during office hours against enemies located thousands of kilometers away. These remarks don't want to be derogatory for the servicemen involved in such activities which require strong technical, juridical and psychological preparation but are not comparable to the risks and stress of direct combat.

Addressing and improving the present rules governing cyber warfare is not an easy task. Most countries have already established cyber structures integrated in their military chains of command and fully dedicated them to cyber defense and to cyber offense. In many ways the spirit is already out of the bottle; "militarization" is a fait accompli: all modern armed forces rely heavily on computers and would be totally paralyzed should they renounce them.

4 Cyber Strategy in Different Countries and Organizations

In the US, President Obama, already in 2009, declared that digital infrastructure was a "strategic national asset" and in May 2010 the Pentagon set up its Cyber Command (USCYBERCOM) in Fort Meade Maryland. According to the 2015 US National Security Strategy "cybersecurity requires that long-standing norms of international behaviour—to include protection of intellectual property, online freedom, and respect for civilian infrastructure" should be respected.

The NATO strategic concept and subsequent NATO summit and ministerial declarations recognized the urgent task of protecting the Alliance's information and communication systems and the necessity to integrate cyber defense into NATO's defense planning. These concepts were further elaborated at the recent NATO summit in Warsaw in July 2016 where it was decided that cyber defence will continue to be integrated into operational planning and Alliance operations and missions. NATO commitment to act in accordance with international law, including the UN Charter, international humanitarian law, and human rights law, was reaffirmed. NATO also addressed the question of an ad hoc new legislation indicating its preference for "voluntary international norms of responsible state behaviour and confidence-building measures regarding cyberspace" (2016).

The original response of the European Union (EU) to the development of a cyber threat was prevalently of a civilian nature. It took place in 2013 through the establishment of the EU Agency for Network and Information Security (ENISA). Globally, the EU strives "for an open and secure cyber realm, in which cyber issues are firmly anchored within the framework of human rights, rule of law and international law". The cyber threat was addressed in the new EU Security Strategy, which was announced in June 2016 by the EU High Representative for Foreign and Security Policy, Federica Mogherini. The language adopted contains substantial policy indications within the EU as well as with external partners. Reference to norms which should regulate this new chapter of international security will hopefully be developed in greater detail in the future.

Cyberwarfare is not a prerogative of the Euro/Atlantic region. Other areas and countries such as Russia, China, India, North Korea, Israel and many others are active in developing military capabilities in this field. In general, countries having acquired such capabilities are reluctant to negotiate international norms which could compromise the advantage they have reached.

So far the world has not experienced cyber attacks of the scale and effect allowing the international use of force. Even in the most acute known cyber episodes, such as the use of Stuxnet which temporarily paralyzed an Iranian nuclear enrichment installation, and the cyber attack against Estonia which in 2008 swamped the websites of Estonia including parliament, ministries, newspapers. Neither Iran nor Estonia invoked Chap VII or art 51 of the UN Charter nor did Estonia, as member both of NATO and the EU, invoke the NATO and EU defence clauses (NATO Art. 5 of the North Atlantic treaty; EU art 42 and 222 of the Lisbon Treaty 2010).

5 Cyberwarfare Legal Status

Cyberwarfare has not yet acquired a legal status of its own in spite of the fact that after land, sea, air and outer space it has become the new fifth domain in which states can confront each other militarily (Trezza 2016). All these domains have been subject, throughout the years to some kind of normative regulation. Even in the case of Outer Space, which preceded cyber warfare as the latest addition to new warfare theaters, an international treaty was finalized only ten years after the launch of the first Sputnik.

Nothing similar happened in the cyber world. As of this moment, States have not even started to negotiate any sort international regulation. This does not mean that the issue has not been addressed. Since 1998 there have been annual reports by the UN Secretary-General to the General Assembly containing the voluntary information of UN Member States on their cyber activities. Such reports have been submitted by a limited number of Member States.

The attention of the UN became more focused when a Group of Governmental Experts (UN) was established on occasion of the Geneva Conference on Disarmament in 2012. Its mandate was to examine the existing and potential threats from the cyber-sphere an possible cooperative measures to address them. This group has now been renewed four times and its main focus has been the establishment of voluntary non-binding norms and confidence building measures, international cooperation and capacity building. In the latest edition of the GGE report a clear priority was given to the effects of cyber attacks against critical infrastructures and terrorism.

6 The Tallinn Manual

Cooperative Cyber Center of Excellence in the Estonian capital Tallinn in order to enhance NATO's cyber defense capability. The center produced the Tallinn Manual on the International Law applicable to cyber warfare, authored by 20 leading international legal experts. In spite of the fact that the Manual has no international legal value, it has the merit of having established some fundamental principles.

According to the Manual, the general principles of international humanitarian law apply to cyberwarfare both in the conduct of war (*jus in bello*) and during the process leading to a war (*jus ad bellum*). This implies that the prohibition of the use or threat use of force enshrined in article 2 of the UN Charter also applies to cyberwarfare. This also indicates that cyber weapons can only be used for self defense or with the authorization of the UN Security Council under chapter VII of the UN Charter.

The Tallinn Manual, defining of the use of force, also establishes the principle of equivalence between a cyber attack and a kynetic attack:

> A cyber operation constitutes a use of force when its scale and its effects are comparable to non cyber operations rising to the level of a use of force (Tallinn Manual Rule 11).

It is not questionable that in a war situation (*jus in bello*), the norms applicable to cyber are the same as those applicable to the other types of weapons and in particular the International humanitarian law. Therefore general principles of necessity, of proportionality between offense and response, the protection of civilians, the clear military advantage expected from an operation are also applicable to cyber weapons.

7 What About Preventive Use of the Cyber Power?

The general norms mentioned so far apply to an already existing conflict situation. Nothing prevents the international community from considering cyber warfare also in a preventive mode, and to craft cyber-specific rules prohibiting cyber instruments capable of provoking catastrophic consequences. In an ideal world, new ad hoc norms should be devised to deal with the strategic and humanitarian implications of cyber war and cyber weapons. Initiatives in this field have been taken by academics and experts, but the bold step of passing from academic and political deliberations to full-fledged negotiations has not yet been taken. As in other similar scenarios, countries having reached a technological advantage are reluctant to tie their hands in a negotiation. But unless codes of conduct, limitations and reductions are negotiated, a destabilizing arms race in cyber power is unavoidable. Every state will pursue what it believes is its immediate interest: without considering the general long term implications.

What are the options today? A general prohibition of possession and use of cyber offensive capabilities would ideally be the preferable solution. We are nowhere close to such a solution. A prohibition limited to an offensive use of cyber weapons, meaning that countries would be allowed to possess but not to use such a type of weapon, could be a more realistic option. However recourse to such a type of arrangement is normally motivated by humanitarian reasons, i.e. to avoid causing unacceptable harm to the civilian population and to reduce the suffering of combatants. In past experiences the humanitarian approach was closely connected to "graphic" episodes of cruelty which had an impact on public opinion and mobilized the NGOs. The humanitarian consequences of a malevolent use of the cyber system as a weapon have not been fully evaluated so far, and thus the establishment of a coalition of likeminded countries spearheading the negotiation of an international ban, as was the case for anti-personnel landmines and cluster munitions, has not been achieved yet.

8 Conclusion

A mechanism specifically dedicated to the risks connected with a cyber warfare still has to be invented. In the meantime, as a first step, preliminary measures of transparency and confidence building, as indicated by the UN GGE must be set up.

> Confidence-building measures increase cooperation and transparency and reduce the risk of conflict. The Group identified a number of voluntary confidence-building measures to increase transparency and suggested that States consider additional ones to strengthen cooperation. The Group called for regular dialogue with broad participation under the auspices of the United Nations and through bilateral, regional and multilateral forums (UN General Assembly 2015b, Para five).

A clearer overall picture of the risks connected with cyber warfare is also necessary. The work done by UN experts in this field is encouraging. An uncontrolled cyber incident could act as a shock absorber to prevent a conflict from escalating. Nonetheless, it could also become a trigger for a wider, global, weapons confrontation. The international community cannot live with such uncertainty nor run such risks. The problem of how to address the security implications of a cyber world will be with us for the years to come. Better to address the issue of a cyber escalation in a preventive mode rather than subsequently to a possible cyber confrontation.

References

NATO (2010) Lisbon Summit Declaration. Retrieved from: http://www.nato.int/cps/en/natohq/official_texts_68828.htm
NATO (2016). Warsaw Summit Communiqué issued by the heads of state and government participating in the meeting of the North Atlantic Council in Warsaw. Retrieved from: 70/71. http://www.nato.int/cps/en/natohq/events_132023.htm
Trezza C (2016) Land, sea, air, space and cyber warfare: diplomatic approaches. In: García-Segura LA, Martin Ramírez J (eds) Mapping the cyberspace. An emerging priority challenge. Universidad Antonio de Nebrija, Madrid, p 27
UN General Assembly (2015a) Creation of a Group of Governmental Experts (GGE) for 2016–2017. Retrieved from: https://www.lawfareblog.com/cybersecurity-un-another-year-another-gge
UN General Assembly (2015b) Group of governmental experts on developments in the field of information and telecommunications in the context of international security. UN General Assembly document A//70/174, 22 July 2015
UN General Assembly (2015c) Resolution 70/237, 23 December 2015, welcoming the outcome of the 2014/2015 GGE and requiring the secretary-general to establish a new GGE that would report to the General Assembly in 2017
UNSC (2004) Resolution 1540 adopted by the UN Security Council at its 4956th meeting, on 28 April 2004

Author Biography

Carlo Trezza was Italy's Ambassador for Disarmament and Non Proliferation and Ambassador to the Republic of Korea. He chaired the Conference on Disarmament in Geneva, the UN Secretary General's Advisory Board for Disarmament Affairs and the Missile Technology Control Regime (MTCR).

Security of Cyber-Space in Nuclear Facilities

Ali Asghar Soltanieh and Hamid Esmailbagi

Abstract Fast efficient promotion of "Information Technology" (IT), expansion of visual space, vast application of internet all over the world have contributed a lot to the "man to man relation" contributing to sustainable development in the 21st century. However, the instrumental malicious use of such an advanced technology has created a serious global concern. Serious cyber-attacks occurred in all over world have given warnings to public and decision maker. It seems no one is immune. Thus it needs collective mobilized efforts to combat such threat. Among the cyber-attacks, the most serious and worrisome is the attacks against nuclear facilities since it shall have radiological consequences for the mankind and the environment. It is no more speculation but a bitter reality. Iranian nuclear facilities were attacked, by so called "Stuxnet", by those who could not tolerate sustainable development through peaceful uses of nuclear energy in Iran. Though the top eminent IT expert of Iran were able to neutralize the attack and prevent the planned impacts, and the fact that they are all alarmed and prepare to defend, for the sake of all people of the world, the peace loving experts have to work together at this time juncture before it is too late. Convening such important conferences, facilitating exchange of information and experiences would surely contribute to the common goal. The Cyber security fits in the domain of nuclear security, which has got more attention nowadays. The International Atomic Energy Agency (IAEA) has taken

Former Representative of the Islamic Republic of Iran to the IAEA.

Submitted: 3.7.16; Accepted: 6.12.16.

A.A. Soltanieh (✉) · H. Esmailbagi
Atomic Energy Organization of Iran, Tehran, Iran
e-mail: energyhasteh@gmail.com

H. Esmailbagi
e-mail: hesmailbegi@aeoi.org.ir

J.M. Ramírez and L.A. García-Segura (eds.), *Cyberspace*, Advanced Sciences and Technologies for Security Applications, DOI 10.1007/978-3-319-54975-0_17

some steps, such as technical meetings on promotion of industrial security, SCADA, but has to further promote its activities on the protection of nuclear facilities by armed attacks as well as cyber-attacks.

Keywords Cyber space · Cyber ware · Stuxnet malware · Nitro Zeus

1 Introduction

The cyber-attack against nuclear facilities within the fuel cycle, specifically the enrichment is no more an anticipated probable scenario, but a reality. The cyber-attack using the so called Stuxnet, under the "Nito Zeus Plan" against Iranian enrichment facilities, was aimed at paralyzing the peaceful activities which are under the full scope safeguards of the International Atomic Energy Agency (IAEA).

Since any attack against nuclear facilities could release radioactive particles which do not recognized international borders, therefore it is a global concern.

It has to be recalled that the Ministerial Declaration, at the IAEA in 2013, emphasized:

> The responsibility for nuclear security within a state rests entirely with that state". It also stressed the importance of international cooperation in supporting states, upon their request, to fulfill their nuclear security responsibilities and obligations and emphasize the need for the involvement of all member states of the agency in its nuclear security – related activities and initiatives. (IAEA in 2013)

The IAEA "Technical Meeting on Conducting Cyber Threat Assessments at Nuclear Facilities" elaborated thoroughly the following main areas:

1. Methodologies for conducting "Cyber Threat Assessments" that can be applied within nuclear security.
2. Methodologies and good practices in computer security at nuclear facilities for guarding against the cyber insider threat.
3. Methodologies and good practices for integration cyber threat considerations into the design basis threat (DBT) (IAEA 2016a, b).

Convening international conferences on Critical Infrastructure Protection (CIP) during recent years indicates that the intellectual communities have noticed the importance of the issue. However, the urgency of the matter, as well as the necessity of collective work, is not felt yet. Researches are being performed in different countries in confidential manner without exchange of technological information. We have still a long way to go to reach to the point of the convergence of scientific works for such common objective, combating the common threat. Such international conferences may increase the awareness of decision makers of the gravity of the threats to the extent that they are convinced to compromise the monopoly for the sake of national and internal security.

From the economic point of view, the cyber-crime direct and direct damage to the world economic is estimated about 400 billion dollars which shall drastically increase in twenty years, including the huge investment that research centers of giant companies have so far invested up to 1.5 billion dollars.

Apart from the economic concerns, the violation of privacy of citizens of the world by cyber–crimes is matter of serious concern and an obstacle for effective fight against cybercrime.

There is a paradox; while the governments of different countries witness the continuous threats and attacks against their countries, specifically their citizens, however some of them do strongly support the hackers and cyber-criminals, with the assumption that they are serving them to combat against the adversaries. They lose sight of the fact the knowledge could not be contained and the experts of the other side can have the access to, depending on the price, and soon they will use against them.

2 The Nature of Cyber Space

In cyber-space the identity and the location of the players are kept secret, anonymous names are usually used. This makes the legally process against the cyber-criminal very difficult.

Considering the fact that security has three main specificities—Accessibility, Integrity, and Confidentiality, the materialization of "security of cyberspace" is very difficult. The cyber is moving very fast. The former head of CIA, during President Bush Administration, confessed that the cyber was moving so fast that we were always a step behind it on the politics.

Though cyber-attack is different from the attack by atomic weapons, however both of them have one thing in common: the total destructions of infrastructure, material, financial, and moral damages to the people of the world.

3 Cyber Attacks Against Iran's Nuclear Facilities

1. FLAME, developed in 2007, was used in attack in May 2012
2. DUQU, developed in 2007, was used in attack in September 2011
3. STUXNET developed in 2009, was used in attack in February 2010.

4 Impacts of Stuxnet Virus on the Strategic Nuclear Program of Iran

According to a report of the IAEA on 18 February 2010, there were indications of virus in the Iranian nuclear facilities. The US International Institute for Strategic Studies (IISS), known for its close collaboration with intelligence services on Iran's nuclear activities, reported that over 1000 centrifuge machines were affected, thus removed and replaced by new ones. However, the same institute confessed that attackers failed in their attempt to destroy all Iranian centrifuges. According to the research made by Symantec Company, the stuxnet virus attacked the controlling systems of the motors of the centrifuge machines and their speeds.

5 Nitro Zeus Plot Against Iran's Nuclear Activities

The plan "Nitro Zeus" seems to be designed by the US Defense, Pentagon, for cyber-attack against Iranian telecommunication, electricity and air defense, as well as nuclear facilities, in case the negotiation would failed. The designed and preparation of this plan for a comprehensive cyber-attack was made during 2009–2010, upon the instruction of the US president to General John Allen. This was however an attempt, as was said, to replace the Israeli attack which could have unexpected consequences for American security and interest in the region.

The industrial sabotage providing Iran with conventional equipment used in any industry with prefabricated electronic malfunctions was discovered by Iranian experts at a nuclear site. The senior author of the present chapter, as the representative of Iran to the IAEA, raised this irresponsible dangerous industrial sabotage at the General Conference. As the result of active diplomacy, this issue was incorporated in the nuclear security resolution.

6 Threat Is Global!!

This threat is not merely the concern of Iranian nuclear industry. The South Korea's state-run nuclear operator was also the subject of a cyber-attack in December 2014, which saw the theft of sensitive information, including the blueprints of at least two nuclear reactors and electrical flow charts.

The authorities of Belgium and other European Countries expressed fears of for Nuclear Power Plant in Belgium (Rubin and Schreuer 2016). And thousands of U.S businesses and other institutions have been besieged by cyber-attacks in recent years.

It is worth to mention that Lt.Gen.Edward Cardon, the head of Joint Task Force Ares. Of the United States declared: "We want to take cyber out of the shadow, where people think we're doing something malicious or spooky, and treat it like we do our operations in other domains".

In 2012 the President Obama signed the Presidential Policy Directive 20, which stipulated that offensive cyber-operations would require presidential approval if they could result in loss of life, serious levels of retaliation, damage to property, adverse foreign policy consequences or economic impact.

However on July 26, 2016, Mr. Obama approved a new Presidential directive that attempts to further define cyber-attacks and what constitutes a "significant cyber incident" as well as how to respond.

Therefore U.S is pursuing more aggressive and offensive policy in cyber ware.

The Democratic National Committee (DNC) hack was an excuse for the United States to take a more robust and open debate about cyber-conflict than is has had to date.

Offensive cyber-capabilities have been largely hidden. The United States has been loath to admit that it has cyber weapons or has used them, such as the secret stuxnet computer worm used to attack Iran's nuclear program.

Needless to say that during the Cold War, the nuclear arms race involved many secrets, but policy and doctrine about the atomic bomb were widely debated.

Some officials of the United States claim that nuclear weapons employment policy has been kept classified but it issued a declaratory policy, open for all to see. Thus U.S. Policy makers are forced to use a simple declaratory policy for cyber (Washington Post 2016)

Therefore, no one is immune. We got to work together by transparent responsible exchange of information based on the lessons learned.

As a useful contribution, one could welcome the EU support to the IAEA on:

- Awareness of and Strengthening states' Response and Resilience to Cyber-Crime impacting Nuclear Security.
- Developing Additional Laboratory Capacity to support Evaluation of Industrial Control and Electronic System Level Technologies used to identify Vulnerabilities to Nuclear—related Cyber-Crime, and Exploiting and Increasing Awareness of such issues, including through Participation in Regional Exchange's, and the Utilization of Compensatory or Remediation Measures (2014).

In the course of discussions at the High–Level Event on "Nuclear Security Summit 2016 and Beyond: The role of Training and Support Centers and Centers of Excellence", held in Bologna, Italy on 7–8 May 2015 emphasis was made on:

1 Share experience on nuclear security.
2. R&D among national institutes in order to ensure scientific advancements and continuous engagement of the scientific communities (e.g. detection, models/simulations, forensic, cyber-security).

   1. Providing technical support for decision—makers.

In this context, Seven Key elements for the sustainability of NSSC/COE, rooted at national, regional and international level, were highlighted:

1. Commitment
2. Integration
3. Building Trust
4. Synergies
5. IAEA Role
6. International cooperation
7. Monitoring sustainability.

7 Multidimensional Phenomena

There is not a "one-size-fits-all" solution to counter cyber incidents and the older approach is outmoded and not sufficient to deal with today's threats to ICT systems. Therefore:

1. A mix of technical, organizational, cultural-focused solutions is needed in order to enhance "resilience" of the ICT system and web
2. Inbuilt "security-by-design" through the whole system
3. Strengthen a "cyber-security culture" at each level of multi-layered complex infrastructure.

8 Importance of a Cyber Security Culture

Cyber security culture involves perception, culture background, latent knowledge and experience beyond the scientific and technological assets. Any operator/organization dealing with cyber security should:

1. Enable compliance with major cyber security related standards
2. Validate cyber security arrangements in external suppliers
3. Provide a foundation for cyber risk assessment
4. Form a basis for policies, standards and procedures
5. Raise cyber security awareness
6. Develop or improve cyber security in response to changing threats
7. Improve resilience against the ever changing threat landscape.

How can a "Hacker" find loop holes? Who support these "Hackers"?

The rate of successful breaches continues to rise. Today's hackers are well funded, often state-sponsored and more sophisticated. These types of attackers easily bypass and defeat traditional security solutions and are forcing organizations to look at new techniques and search for new technologies to counter this growing threat. The threats observed are including but not limited to the following:

1. Unauthorized access and exploitation of Internet facing ICS/Supervisory Control and Data Acquisition (SCADA) devices
2. Exploitation of Zero-Days (ZDEs) vulnerabilities in control system devices and software
3. SQL injection via exploitation of web application vulnerabilities
4. Network scanning and probing
5. Advanced persistent threats (APT)
6. Lateral movement between network zones
7. Targeted spear-phishing campaigns.

The resolution GC (57)/RES/10 of the 57th General Conference of the IAEA noted the IAEA's efforts to raise awareness of the threat of cyber-attacks and their potential impact on nuclear security, and encouraged the IAEA to make further efforts to improve international cooperation in this regard. However, words have not been translated into action. The Agency has to bring top experts from countries with nuclear activities with the objective of informing their experiences and lessons learned in order to prepare series of preventive measures guidelines as well as emergency safety measure in case of cyber-attack against nuclear facilities. Main elements of the Guidance: Computer-based systems play an ever expanding role in nuclear security. These roles include sensitive information management, nuclear safety systems, physical protection systems, and nuclear material accountancy and control at nuclear facilities and associated operations. Computer-based systems are increasingly included in new designs and may be introduced to existing facilities during modernization, or to increase productivity (IAEA 2016a, b).

Computers and computing systems are used in all aspects of nuclear material processing. Therefore there are potential targets:

1. Seeks to create awareness of the importance of incorporating computer security as a fundamental part of the overall security plan for nuclear facilities.
2. Provides guidance to nuclear facilities on evaluating existing programs, assessing critical digital assets and identifying appropriate risk reduction measures.
3. Provides guidelines to personnel designing, implementing, and managing Instrumentation and Control (I&C) and Information systems and networks at nuclear facilities.
4. Establishment of cyber -security unit in nuclear facilities consist of expert from departments of IT, nuclear safety and radiation protection, security, and safeguards, due to their intertwined relation.
5. Specific urgent attention has to paid to the cyber-attack against nuclear installations before the world would face with radiological consequences.

9 Security of Cyber Space Vis-à-Vis Nuclear Security

The term nuclear security covers all security aspects of nuclear facilities or activities. At the same time, one has to notice that nuclear security is intertwined with nuclear safety and nuclear safeguards. Each of these three, so called 3S, are related to specific international norms, regulations, conventions.

Nuclear security covers, any deliberate act which endangers the security of a nuclear facility or activities. Therefore, military attack, terrorist attack, cyber-attack by a terrorist or a state, industrial sabotage, and the assassination of nuclear scientist or personnel… all of them are related to nuclear security.

10 Iran's International Initiative

Pursuant to the military attack by Israeli regime against Iraqi reactor in 1981, followed by a military attack against Bushehr Nuclear Power Plant of Iran by the Iraki Sadam regime, the senior author proposed on behalf of the Islamic Republic of Iran a draft resolution, which was passed in 1990. According to the mentioned resolution, Res/533, any armed attack and threat of attack against a nuclear installation during operation and during construction constitutes violation of the UN Charter, IAEA Statute, and the international law. In such cases, the United Nation Security Council has to act immediately. After almost two decades, when the threat of attack by Israeli regime against Iranian nuclear facilities were augmented, the author, on behalf of the Islamic Republic of Iran proposed a text to with the same content, to the IAEA General Conference in 2009, which was adopted, by consensus, as the Presidential Statement. The following year at the NPT Review Conference a similar text was proposed by Iran and it was also adopted by consensus. Therefore, there is an internationally accepted document in this regard paving the way for next step to turn it to a legally binding instrument.

Considering the fact that the nuclear security is a global concern, the Nuclear Security Summits held in Washington DC (2010), Seoul (2012), The Hague (2014), and the last one in Washington again (2016) with participation of about 50 selected countries. is an exclusive approach which dooms to failure. In order to put things on the right track Iran and other like-minded countries proposed in the IAEA to hold high level conference open to all Member States. The first Ministerial Conference was held in 2013 and t he 2nd is planned for December 2016.

It is highly recommended that a clear message on the security of cyber space of nuclear facilities be sent to the international community by the Madrid 2016 CICA International Conference on Mapping the Cyberspace..

11 Conclusion

1. The interconnection of various institutions such as digital infrastructure organizations, business community, law enforcement, safety regulatory bodies, and NGOs, all have to collaborate closely in order to effectively combat the cyber-attacks.
2. Any cyber-attack against nuclear facilities has radiological consequences with the transfer of radioactive material beyond international borders, as happened in the case of nuclear accidents (Chernobyl and Fukushima). Therefore, it is a global threat, that requires prompt collective international actions both for prevention and for emergency assistance.

References

IAEA (2013) Computer security of nuclear facilities, IAEA guideline. Retrieved from: http://www-pub.iaea.org/mtcd/publications/pdf/pub1527_web.pdf

IAEA (2016a) Computer security techniques for nuclear facilities IAEA guideline, 2016. Retrieved from: http://www-ns.iaea.org/security/nss-publications.asp

IAEA (2016b) Technical meeting on conducting cyber threat assessments at nuclear facilities, IAEA, 9–12 February 2016. Retrieved from: http://www-ns.iaea.org/security/nss-publications.asp

Rubin AJ, Schreuer M (2016) Belgium fears nuclear plants are vulnerable. Retrieved from: http://www.nytimes.com/2016/03/26/world/europe/belgium-fears-nuclear-plants-are-vulnerable.html?_r=0

Washington Post (2016) The post's view, bring cyber warfare further out of the shadows, 1 August 2016. Retrieved from: https://www.washingtonpost.com/opinions/bring-cyberwarfare-further-out-of-the-shadows/2016/08/01/abc20800-4def-11e6-aa14-e0c1087f7583_story.html?utm_term=.4c877c44f123

Author Biographies

Ambassador Ali Soltanieh (Iran) is Advisor to Vice President and Head of Atomic Energy Organization of Iran, and Senior Researcher at the Institute of Political International Studies (IPIS). Trained as a nuclear physicist, Ambassador Soltanieh has been involved in scientific and diplomatic activities, as a nuclear physicist and senior diplomat, during the last thirty five years. He has been involved in the WMD non-proliferation and disarmament, including all together 12 years as Ambassador and representative to the IAEA since 1982, three years as Secretary of National Authority for Chemical Weapons Convention and three years as Chief Negotiator for Biological Weapons Convention. He has followed the issue of Nuclear Weapon Free Zone in the Middle East as well as nuclear safety, nuclear security and safeguards for the last three decades. Ambassador Soltanieh has participated, n the capacity of special envoy, delegate, chief negotiator, and invited speaker, in over 180 international events on nuclear science and technology as well as WMD disarmament and international security, such as NPT, CWC, BWC, CTBT, CCW, and has worked closely with the relevant international scientific and technical organizations such as IAEA, OPCW, and other specialized international organizations such as UNIDO, OPEC, OFID, ILO, IPU, OIC, WMO, TWAS and ICDO. He has had interactions with the known Think-Tanks and NGOs in the world. He has been participating as panelist on WMD at the Pugwash conferences in Hiroshima (2005), Istanbul (2013, on chemical weapons panel), Nagasaki (2015), EU Consortium on Non-proliferation and Disarmament (2011, 2013, and 2015). He has published several papers and books in afore-mentioned areas, and has gotten several National and International Admirations and Medals.

Eng. Hamid Esmailbagi is Section Head of IT, at the Atomic Energy Organization of Iran.

Can Cyber Attacks Prevent Wars?

Gunnar Westberg

Abstract Cyber attacks have in many instances interrupted production facilities, banking, management of public and private enterprises and also interfered with military planning and military exercises. I propose here that cyber attacks can promote conflict resolution and prevent war. The possibilities and limits of this form of "asymmetric warfare to promote peace" are analyzed. Will the method work best to avert an attack by a large state on a small country, or can conflicts between large powers also be delayed and maybe solved by the use of cyber threats? Is secrecy needed and is it possible? Today, when the risk of nuclear war is said to be increasing, it is increasingly important to find non-military ways to defuse and avoid conflicts.

Keywords Preventing war · Conflict resolution · Cyber war · Cyber terrorism · Asymmetric warfare

1 Introduction

Most cyber attacks are intended to obtain information, e.g. from businesses, banks, research facilities or government. However, sometimes the attack aims at disrupting certain activities that are dependent on the function of computer networks. Such offensive cyber attacks, sometimes referred to as cyber terrorism, is a problem which is currently much discussed, as are means to detect and defend against such

---

Former Co-president of International Physicians for the Prevention of Nuclear War.
Submitted: 31.8.16, Admitted: 11.10.16.

G. Westberg (✉)
Goteborg University, Gothenburg, Sweden
e-mail: gunnar.westberg@medic.gu.se

J.M. Ramírez and L.A. García-Segura (eds.), *Cyberspace*, Advanced Sciences and Technologies for Security Applications, DOI 10.1007/978-3-319-54975-0_18

assaults both from individuals, terrorists and states (U.S. Department of Defence 2013; Zetter 2014).

However, I suggest that cyber attacks can also be an alternative to conventional weapons, making their use unnecessary. This paper will discuss if, and in what situations, cyber attacks can prevent a conflict from escalating to a regular war.

2 Scenarios

Scenario 1. How a small country can avert an attack by a large country

Here I describe a possible setting, entirely fictional, intended to illustrate how a small country can avert an attack by a large country, using cyber attacks by proxy.

We assume here that the large Russian minority in Estonia becomes increasingly upset by proposed changes in the legislation regarding the right to become a citizen of Estonia. These groups ask the Russian government to press the Estonian parliament to stop the change of the law. Russian leaders then demand that the controversial legislation is revoked. If not, sanctions will be applied. Finally, the possibility of military "protection of the Russian minority" is hinted.

Suddenly, the stock exchange in Moscow and the electricity distribution in St Petersburg stops functioning for a whole day. Estonian authorities claim that they have nothing to do with this disturbance but cannot exclude that Estonian hackers have been active. The Russian President demands an excuse from the government of Estonia.

A group of Estonians living in other countries sends a secret message to the Russian government declaring that they have computer viruses and "worms" planted in several servers in different countries, that could bring the Russian economy to a standstill. As a warning, sections of the Moscow Metro are suddenly left without electricity for several hours. Estonian journalists report that leaks from the Internet point to a specific radical group of Estonians in Germany as responsible.

As the threat of a major attack on Russia's economy is not public, the Russian leaders can retreat without a loss of face. They ask for an international mediation in the conflict regarding the status of the Russian minority. "Russia always wants good relations with its neighbors and respects international law".

This type of conflict is not uncommon. Often the conflict is left pestering for years and good relations between the countries do not develop. Escalation into a major military conflict can occur because of this conflict alone, or when additional problems arise. In similar conflicts cyber attacks may prevent escalation to an all-out war.

Scenario 2. Can cyber warfare prevent a military conflict between two superpowers?

Here is an attempt to illustrate the idea.
Today both USA and China are rapidly increasing their military presence in parts of the Pacific, particularly in the area referred to as the South China Sea (BBC 2016). There are territorial problems between China and the countries in the region, some of them with military alliances with the USA. There may be oil resources. The waters are of great importance for international trade, especially for China.

In a situation of an escalating confrontation between USA and China in the area, we could imagine that China accesses a selected part of a US computer network, causing e.g. a breakdown of the electrical distribution in parts of California. China may leave a characteristic hallmark on the virus, making it possible to trace the origin of the attack, but could instead choose to have an anonymous virus sent from, e.g., North Korea. The intrusion could be read as a warning to the USA: "If you go further in your military activity we have the means to close down your computer networks and bring your economy to a standstill for weeks." To add emphasis, parts of the US Internet are disrupted repeatedly for a few hours, after anonymous attacks.

The USA might in this situation retaliate in kind with an attack on Chinese computer networks. That would risk an escalating counterstrike from China causing a breakdown of important networks in USA.

If the USA instead chooses a military strike, China may launch a devastating attack on the Internet, which would bring the economic activity in the USA to a halt for days, maybe including television networks. Destructive and dangerous interference with the function of e.g. nuclear power plants or infrastructure could threaten. Considering these possibilities, negotiations would seem to be the best way forward for both parties.

In both scenarios, it would be a great advantage in many cases if the threats are made in secret. Prestige is important here, just as in ordinary military conflicts. Secrecy makes "face-saving" possible.

3 Discussion

A few questions should be discussed:

1. Can cyber strikes have such severe consequences that the threat of their use can motivate a superpower to refrain from a strong military action to defend important national interests? That seems to be the case: The report from the US Department of Defence Science Board (2013) discusses cyber attacks that are so devastating that they are compared to nuclear attacks (Colby 2013). To avoid such a cyber strike, negotiations should certainly be tried first.

2. Can and should cyber attacks remain anonymous and secret? The origin of sophisticated attacks is often not possible to trace (Lipson 2002; Hackmageddon 2016). Several attacks on national or defense networks happen every day and the attackers can often not be identified. In many situations, it may lay in interest of the attacked country to remain ambiguous about its suspicions of the source of the intrusion. This would give time to consider non-military means to solve the conflict. If the source is not identified, it may be easier to back down in secret negotiations.
3. Is there not sufficient redundancy in the computer networks in big countries, today or a few years in the future, to make a serious disruption unlikely? The report mentioned above (U.S. Department of Defence 2013) states that "the United States cannot be confident that our critical Information Technology (IT) systems will work under attack from a sophisticated and well-resourced opponent…"

In this report, we also learn that the Department of Defence's own "Red Gangs"—hackers checking the security of the systems- are quite successful in breaking into and disturbing important networks.

Much work is going on to strengthen the defence of the networks and to improve the capacity to identify and stop intruders and hackers form disrupting the function. However, when defence is strengthened, so is the offensive capacity.

The use of cyber attacks as a form of asymmetric warfare intended to prevent a military attack has never been tried, as far as the author knows. However, more and more countries become technically sophisticated and able to execute measured cyber attacks successfully, and even secretly. We know that attacks are carried out by countries against the networks of other countries, including the military networks. But in general, what we have seen is just tests, sometimes intended to indicate their capacity. I fear that the worse is yet to come.

A word should also be given regarding the "hacktivist" movement which calls itself "Anonymous". This is not an organization, because it has no leaders who can be identified and no agreed rules. "Anonymous" could be referred to as a brand name. One common purpose seems to be to defend the independence of the Internet, but also freedom of speech and political freedom in general. The method used mostly is Denial of service to the websites of government agencies or corporations. An impressive list of actions is given in a Wikipedia report (2016).

"Anonymous" has been successful in raising awareness in the general public but it is difficult to prove that the actions of the group has in the end caused change. At present it does not seem likely than any citizen group will be able to influence governmental decisions regarding war or peace

4 Conclusion

This chapter proposes that we should start a discussion regarding the possibilities of a new way to prevent wars, namely by using cyber attacks.

A state that feels threatened by another state could attack through secret servers, maybe located in one or more other countries. Such an attack should, to be effective, result in the disruption of one or several important functions in society.

The defending state may have great resources in order to protect its society, by increased redundancy and other means. However, in the opinion of the US department of Defence, quoted here, the protection may be insufficient. A small state with sophisticated technology will quite possibly be able to cause severe damage.

In many situations the attacked country will realize that a military response could result in similar or worse retribution in the form of crippling cyber attacks. Negotiations should be preferred.

If the nature and origin of the cyber threat can be kept secret, neither side would lose face.

References

Anonymous (2016) Timeline of events associated with Anonymous. Wikipedia. Retrieved from: https://en.wikipedia.org/wiki/Timeline_of_events_associated_with_Anonymous

BBC News (2016) Why is the South China sea contentious? Retrieved from: http://www.bbc.com/news/world-asia-pacific-13748349

Colby E (2013) Cyberwar and the nuclear option. Retrieved from: http://nationalinterest.org/commentary/cyberwar-the-nuclear-option-8638

Lipson H (2002) Tracking and tracing cyber-attacks: technical challenges and global policy issues. Retrieved from: http://oai.dtic.mil/oai/oai?verb=getRecord&metadataPrefix=html&identifier=ADA408853

Hackmageddon (2016). Information security timelines and statistics 2016. Retrieved from: http://www.hackmageddon.com/2015/08/18/1-15-august-2015-cyber-attacks-timeline/

U.S. Department of Defence (2013) Defence Science Board: resilient military systems. Advanced cyber threat. Retrieved from: http://www.acq.osd.mil/dsb/reports/ResilientMilitarySystems.CyberThreat.pdf

Zetter K (2014) Countdown to Zero Day: stuxnet and the launch of the world's first digital weapon. Broadway books

Author Biography

Dr. Westberg is Professor of Medicine, emeritus, Co-president of International Physicians for the Prevention of Nuclear War, (2006–2008) and member of the Sahlgrenska Academy of Medicine. Goteborg University. Gothenburg, Sweden.

Erratum to: Cyberspace

J. Martín Ramírez and Luis A. García-Segura

Erratum to:
J.M. Ramírez and L.A. García-Segura (eds.), *Cyberspace*, Advanced Sciences and Technologies for Security Applications, DOI 10.1007/978-3-319-54975-0

In the original version of the book, the missing Dedication page in frontmatter has to be added before Prologue. The erratum book has been updated with the change.

The updated online version of this book can be found at
http://dx.doi.org/10.1007/978-3-319-54975-0

J.M. Ramírez and L.A. García-Segura (eds.), *Cyberspace*, Advanced Sciences and Technologies for Security Applications, DOI 10.1007/978-3-319-54975-0_19

Epilogue

In the history of mankind, particularly the history of science, man often faced new and potentially dangerous phenomena that later became familiar and proved useful, while still retaining their danger potential.

Cyberspace started as an extremely simple, but enormously clever, solution to a very down-to-earth problem, just aimed at saving time and money. Today it has become a monster network of ever growing, immense possibilities. In just the last three decades, it has had enormous consequences for the planet and its inhabitants. There are currently no signs that Cyberspace's power to provoke change, both good and bad, will fade anytime soon.

Some of the most compelling positive changes associated to Cyberspace today have to do with the digital revolution as a catalyst to accelerate the social and economic transformations needed in developing countries.

Nonetheless, the negative changes in and around Cyberspace threaten to undermine any positive gains for society. The use of this medium as an almost unlimited anonymous platform for conducting criminal activities is forcing public and private enterprises to work together in order to design proper legal and functional frameworks. These frameworks must deeply respect the rights linked to freedom, but also to cyber-freedom: the power to decide on existing personal data.

Although we know that the network is controlled by human beings anywhere, we need to know who they are, and in some cases what values move them. This is especially necessary due to the Cyberwarfare aspect of the Internet, where worst case scenarios include cyber-attacks by a variety of state and non-state actors that have the audacity to even target nuclear facilities.

Society must face these challenges counting on the participation, dialogue and agreement of not only the usual multilateral international forums, but also the new emerging social network tools and mechanisms, where the majority of the youth population worldwide are voicing their opinions.

Luis A. García-Segura
Universidad Antonio de Nebrija
Madrid, Spain

J.M. Ramírez and L.A. García-Segura (eds.), *Cyberspace*, Advanced Sciences and Technologies for Security Applications, DOI 10.1007/978-3-319-54975-0

Printed by Printforce, the Netherlands